BARRON'S

Biology
Practice
Plus

400+ ONLINE QUESTIONS

WITH QUICK REVIEW GUIDE

DEBORAH T. GOLDBERG, M.S. AND MARISA ABRAMS, M.S.

Dedication

I wish to dedicate this book to—
—My wonderful children, Michael and Sara Boilen, who I love and cherish

Acknowledgments

I wish to thank—
—My husband, Howard Blue, for his constant love and support

I wish to thank—
—My husband, Sheldon, for his constant and unwavering love, encouragement, and support
—My parents for their support of my pursuit of a career in science and education

Published by Kaplan North America, LLC, dba Barron's Educational Series
1515 W Cypress Creek Road
Fort Lauderdale, FL 33309
www.barronseduc.com

ISBN: 978-1-5062-8148-3

10 9 8 7 6 5 4 3 2 1

Kaplan North America, LLC, dba Barron's Educational Series print books are available at special quantity
discounts to use for sales promotions, employee premiums, or educational purposes. For more information or
to purchase books, please call the Simon & Schuster special sales department at 866-506-1949.

About the Authors

Deborah Goldberg, M.S.

Deborah Goldberg, M.S. taught AP Biology and Chemistry at Lawrence High School on Long Island, New York for more than 20 years. She also spent 14 years completing research as an electron microscopist at New York University Medical Center and New York Medical College.

Marisa Abrams, M.S.

Marisa Abrams, M.S. has taught AP Biology, Biology, and Chemistry for 19 years. She has attended the AP Biology Read (as both a reader and a Table Leader), and she has worked with ETS and the College Board in various capacities (item writing/reviewing for AP and serving on Biology committees).

CONTENTS

How to Use This Book

Barron's Biology Practice Plus is designed to offer essential review of key topics and loads of online practice to help you excel in biology.

Online Practice

Access more than 400 questions in online quizzes arranged by topic for customized practice! All questions include answer explanations.

To access these questions, see the card at the front of the book.

What Will You Learn in the Book?

Key review and topics are covered so you can study the essentials needed to succeed.

Learning objectives are listed at the start of each chapter. This list of key ideas will help guide your learning and study plan and allow you to easily return to topics that you want to review again.

Tips are given throughout the book to offer helpful notes, reminders, and strategies to improve your learning.

Biochemistry

Learning Objectives

In this chapter, you will learn:

- Basic chemistry—atomic structure
- Bonding
- Hydrophobic and hydrophilic
- Characteristics of water
- pH
- Organic compounds

To understand about living things, you must understand some basic biochemistry. Biochemistry affects every aspect of our lives every day. Sweating cools the skin because of strong hydrogen bonding between water molecules. Your body maintains your blood at one critical pH because of the bicarbonate buffering system. To prevent heart attacks, you must begin with an understanding about the structure of fatty acids. Mad cow disease is caused by a misfolded protein. Here is a review of basic biochemistry.

Basic Chemistry—Atomic Structure

The atom consists of subatomic particles: protons, neutrons, and electrons. Table 1.1 compares these particles.

TABLE 1.1 Subatomic Particles

Subatomic Particle	Charge	Mass in amu	Location
Proton	+1	1	Nucleus
Neutron	0	1	Nucleus
Electron	−1	0	Outside nucleus

1. An atom in the elemental state always has a neutral charge because the number of protons (+) equals the number of electrons (−).

2. Electron configuration is important because it determines how a particular atom will react with atoms of other elements.

3. Electrons in the lowest available energy level are said to be in the **ground state**.

4. When an atom absorbs energy, its electrons move to a higher energy level. The atom is then said to be in the **excited state**. For example, during photosynthesis, chlorophyll molecules absorb light energy, which boosts electrons to higher energy levels. These excited electrons provide the energy to make sugar as they return to their ground state and release the energy they previously absorbed.

> **CAREFUL**
>
> Don't confuse isotope with isomer.

Isotopes are atoms of one element that vary only in the number of neutrons in the nucleus. Chemically, all isotopes of the same element are identical because they have the same number of electrons. For example, carbon-12 and carbon-14 are isotopes of each other and are chemically identical. They both possess 6 protons and 6 electrons. However, carbon-12 has 6 neutrons, while carbon-14 has 8 neutrons. Some isotopes, like carbon-14, are radioactive (called **radioisotopes**). The nuclei of radioisotopes emit particles and decay at a known rate called a **half-life**. Knowing the half-life enables us to measure the age of fossils or to estimate the age of Earth.

Radioisotopes are useful in other ways. Besides measuring the age of fossils, they can be used in medical diagnosis, treatment, and research. For example: radioactive iodine (I-131) can be used both to diagnose and to treat certain diseases of the thyroid gland. In addition, radioactive carbon can be used as a **tracer**, incorporated into molecules of carbon dioxide, and used to track metabolic pathways.

Bonding

A bond is formed when two atomic nuclei attract the same electron(s). Energy is released when a bond is formed. Energy must be supplied or absorbed to break a bond. Atoms bond to achieve stability, to acquire a completed outer shell. There are two main types of bonds, ionic and covalent.

Ionic bonds form when electrons are transferred. An atom that gains electrons becomes an **anion**, which stands for a negative ion. An atom that loses an electron becomes a **cation**, a positive ion. Ions such as Cl^-, Na^+, and Ca^{2+} are necessary for normal cell, tissue, and organ function.

Covalent bonds form when atoms share electrons. The resulting structure is called a **molecule**. A single covalent bond (−) results when two atoms share one pair of electrons. A double covalent bond (=) results when two atoms share two pairs of electrons, and a triple covalent bond (≡) results when two atoms share three pairs of electrons.

There are two types of covalent bonds, nonpolar and polar. This classification is based on whether electrons are shared equally or unequally. See Table 1.2.

TABLE 1.2 The Two Types of Covalent Bonds

Nonpolar Covalent Bond	Polar Covalent Bond
Electrons shared equally	Electrons shared unequally
Formed between any two atoms that are alike or with an electronegative difference of 0	Formed between any two atoms that are different or with an electronegative difference greater than 0
Examples: H_2 (H – H) and O_2 (O = O) nonpolar bond ↑ ↑ nonpolar bond	Examples: CO (C = O) and H_2O (H – O – H) polar bond ↑ ↑ ↑ polar bonds

Intermolecular Attractions

Not only do atoms *within* molecules attract each other, but there are also attractions *between* molecules. There is a variety of these **intermolecular attractions**. You should know three.

Polar-Polar Attraction

When two or more atoms form a bond, the entire resulting molecule is either polar (unbalanced) or nonpolar (balanced). There are stronger attractions between polar molecules than between nonpolar molecules. The negative end of one polar molecule attracts the positive end of another polar molecule. H_2O is a highly polar (unbalanced) molecule. It looks like this:

Hydrogen Bonding

Hydrogen bonding is very important to living things. Hydrogen bonding:

- Keeps the two strands of DNA bonded together, forming a double helix.
- Causes water molecules to stick together and is responsible for many special characteristics about water. See Figure 1.1.

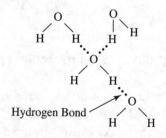

FIGURE 1.1 Hydrogen bonding

Nonpolar Molecules

Only the weakest attractions (van der Waals) exist between nonpolar molecules. An example of a nonpolar molecule is CO_2. It is linear and balanced. It looks like this:

$$O = C = O$$

Hydrophobic and Hydrophilic

Hydrophobic means "water hating" or "repelled by water." **Hydrophilic** means "water loving" or "attracted to water." Substances that are polar will dissolve in water, while substances that are nonpolar will not dissolve in water. Remember: like dissolves like. You are familiar with an open can of soda that quickly loses its fizziness. This is because carbon dioxide, a nonpolar molecule that gives soda pop its fizziness, does not dissolve in water, which is polar. So, gas escapes when you open a can of soda pop and it goes flat.

Lipids, which are nonpolar, are hydrophobic and do not dissolve in water, which is why oil and vinegar salad dressing separates upon standing. Since the plasma membrane is a phospholipid bilayer, only nonpolar substances can readily dissolve through the plasma membrane. Large polar molecules cannot diffuse across a plasma membrane. They can only travel across a membrane through special hydrophilic (protein) channels.

Characteristics of Water

Water is asymmetrical and very polar. It also has strong intermolecular attractions. In addition to polar attractions, water exhibits strong hydrogen bonding, as shown in Figures 1.1 and 1.2. Together, these two forces are responsible for the special characteristics of water that affect life on Earth.

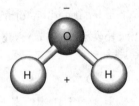

FIGURE 1.2 Water molecule is polar

1. **Water has a high specific heat.** Specific heat is the amount of heat that must be absorbed in order for 1 gram of a substance to change its temperature 1° Celsius. This means that large bodies of water, like oceans, absorb a lot of heat and resist changes in temperature. As a result, they provide a stable environment for the organisms that live in them. Also, coastal areas exhibit relatively little temperature change because the oceans moderate their climates.

2. **Water has a high heat of vaporization.** This means that a relatively great amount of heat is needed to evaporate water. As a result, evaporation of sweat significantly cools the body surface.

3. **Water has high adhesion properties.** Adhesion is the clinging of one substance to another, and it plays an important role in plant survival. Forces of adhesion contribute to capillary action, which helps water flow up from the roots of a plant to the leaves.

4. **Water is the universal solvent.** Because water is a highly polar molecule, it dissolves all polar and ionic substances.

5. **Water exhibits strong cohesion tension.** This means that molecules of water tend to stick to each other. This results in several biological phenomena. Water moves up a tall tree from the roots to the leaves without the expenditure of energy by what is referred to as evapotranspiration. It also results in surface tension that allows insects to walk on water without breaking the surface.

6. **Ice floats because it is less dense than water.** In a deep body of water, floating ice insulates the liquid water below it, allowing life to exist beneath the frozen surface during cold seasons. The fact that ice covers the surface of water in a lake in the cold months and melts in the spring results in a stratification of the lake during the winter and considerable mixing in the spring. In the spring, surface ice melts, becomes denser water, and sinks to the bottom of the lake, causing water to circulate throughout the lake. Oxygen from the surface is returned to the depths, and nutrients released by the activities of bottom-dwelling bacteria are carried to the upper layers of the lake. This cycling of the nutrients in the lake is known as the spring overturn and is necessary to the health of a lake.

pH

pH is a measure of acidity and alkalinity of a solution. It is the negative logarithm of the hydrogen ion concentration in moles per liter. Anything with a pH of less than 7 is acidic, and anything with a pH value greater than 7 is alkaline or basic. A substance with a pH of 7 is neutral; see Figure 1.3.

THIS IS TRICKY

As the concentration of H^+ increases, the pH decreases.

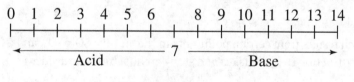

FIGURE 1.3

As shown in Table 1.3, a solution of pH 1 is 10 times more acidic than a solution with a pH of 2 and is 100 times more acidic than a solution with a pH of 3. It is 1,000 times more acidic than a solution with a pH of 4.

TABLE 1.3 pH Compared with Molarity

pH	Concentration of H^+ Ions in Moles per Liter
1	$1 \times 10^{-1} =$ 0.1 molar
2	$1 \times 10^{-2} =$ 0.01 molar
3	$1 \times 10^{-3} =$ 0.001 molar
4	$1 \times 10^{-4} =$ 0.000 1 molar
7	$1 \times 10^{-7} =$ 0.000 000 1 molar
13	$1 \times 10^{-13} =$ 0.000 000 000 000 1 molar

The pH of some common substances is shown in Table 1.4.

TABLE 1.4 pH Values

Substance	pH
Stomach acid	2.0
Orange juice	3.5
Carbonated drinks	3.0
Acid rain	<5.6
Milk	6.5
Human blood	7.4
Seawater	8.5

The internal pH of most living cells is close to 7. Even a slight change can be harmful.

Biological systems regulate their pH through the presence of **buffers**, substances that resist changes in pH. A buffer works by absorbing excess hydrogen ions or donating hydrogen ions when there are too few. The most important buffer in human blood is the **bicarbonate ion** (HCO_3^-).

Acid rain, which results from certain pollutants in the air (like SO_2, SO_4, and CO_2), has caused damage and destruction to many lakes and stone architecture worldwide.

Organic Compounds

Organic compounds are compounds that contain carbon. There are four classes of organic compounds: carbohydrates, lipids, proteins, and nucleic acids.

Carbohydrates

Carbohydrates consist of only three elements: carbon, hydrogen, and oxygen.

- Carbohydrates supply quick energy.
- One gram of any carbohydrate will release 4 calories of heat when burned.
- Dietary sources include rice, pasta, bread, and cookies.
- There are three classes of carbohydrates: monosaccharides, disaccharides, and polysaccharides.

Monosaccharides

Monosaccharides have a chemical formula of $C_nH_{2n}O_n$. A common formula is $C_6H_{12}O_6$. Examples are glucose, galactose, and fructose, which are all **isomers** of each other. Isomers are compounds with the same molecular formula but with different structures. Therefore, isomers have different physical and chemical properties.

The structural formula of glucose is shown in Figure 1.4.

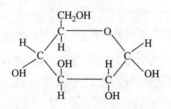

FIGURE 1.4 Monosaccharide—Glucose

Disaccharides

All disaccharides have the chemical formula $C_{12}H_{22}O_{11}$. They consist of two monosaccharides joined by a process known as **dehydration synthesis** or **condensation**.

The structure of the disaccharide, maltose, is shown in Figure 1.5.

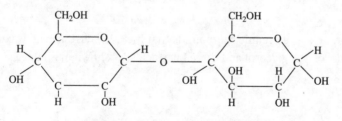

FIGURE 1.5 Disaccharide—Maltose

Table 1.5 shows three dehydration synthesis reactions of monosaccharides. They produce three disaccharides—maltose, lactose, and sucrose—as well as the by-product water.

TABLE 1.5 Dehydration Synthesis of Monosaccharides

Monosaccharide + Monosaccharide → Disaccharide + Water					
$C_6H_{12}O_6$	+	$C_6H_{12}O_6$	→	$C_{12}H_{22}O_{11}$ +	H_2O
Glucose	+	Glucose	→	Maltose +	Water
Glucose	+	Galactose	→	Lactose +	Water
Glucose	+	Fructose	→	Sucrose +	Water

Hydrolysis is the opposite of dehydration synthesis. It is the breakdown of a compound with the addition of water. It is what occurs during **digestion** and is the reverse of dehydration synthesis.

Sucrose + Water → Glucose + Fructose

Polysaccharides

Polysaccharides are polymers of carbohydrates. They form as many monosaccharides are joined together by dehydration synthesis. Table 1.6 shows four important polysaccharides you should know: cellulose, starch, chitin, and glycogen.

TABLE 1.6 Polysaccharides

Found in Plants	
Cellulose	**Starch**
Makes up plant cell walls	The way plants store carbohydrates
Found in Animals	
Chitin	**Glycogen**
Makes up the exoskeleton in arthropods and cell walls in **fungus**	"Animal starch"; in humans, this is stored in the liver and skeletal muscle

Lipids

Lipids are a diverse class of organic compounds that include fats, oils, and waxes. Structurally, most lipids consist of one glycerol and three fatty acids, as shown in Figure 1.6.

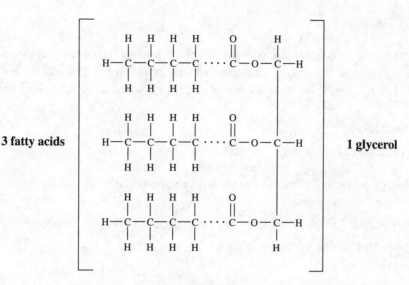

3 fatty acids   **1 glycerol**

FIGURE 1.6

Glycerol is an alcohol whose structure is shown in Figure 1.7.

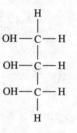

FIGURE 1.7 Glycerol

A fatty acid is a hydrocarbon chain with a carboxyl group at one end. Fatty acids exist in two varieties: **saturated** and **unsaturated**. These are shown in Figure 1.8.

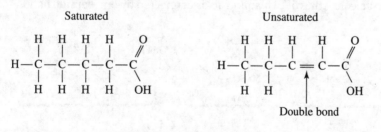

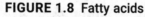

FIGURE 1.8 Fatty acids

Saturated Fats

Saturated fats, with a few exceptions, come from animals. They are solid at room temperature and, when ingested in large quantities, are linked to heart disease. An example of a saturated fat is butter. Saturated fatty acids contain only single bonds between carbon atoms.

Unsaturated Fats

Unsaturated fatty acids are extracted from plants, are liquid at room temperature, and are considered the "good dietary fats." Unsaturated fatty acids have at least one double bond between carbon atoms in the hydrocarbon chain. Thus, they have fewer hydrogen atoms.

Lipid Functions

Lipids serve many functions.

1. **Energy storage:** 1 gram of any lipid will release 9 calories of heat per gram when burned in a calorimeter.

2. **Structural:** Phospholipids are a major component of the cell membrane.

3. **Endocrine:** Some lipids are hormones.

Proteins

- Proteins are **polymers** or **polypeptides** consisting of repeating units called amino acids joined by **peptide bonds**.

- Amino acids consist of a carboxyl group, an amine group, and a variable (R), all attached to a central carbon atom. The R group, or variable, differs with each amino acid.

- Proteins are complex macromolecules and are responsible for growth and repair.

- Dietary sources of proteins include fish, poultry, meat, and certain plants called legumes, like beans and peanuts.

- One gram of protein burned in a calorimeter releases 4 calories of heat.

- Proteins consist of the elements S, P, C, O, H, and N.

- With only 20 different amino acids, cells can build thousands of different proteins.

- Enzymes are examples of proteins.

- Figure 1.9 is a sketch of two amino acids combining to form a **dipeptide**. A dipeptide is a molecule consisting of two amino acids connected by one peptide bond.

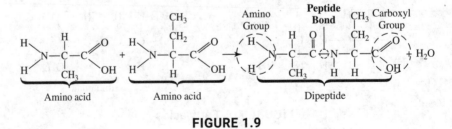

FIGURE 1.9

Protein Structure

Proteins have many functions in living things. They act as enzymes, membrane channels, and hormones, to name a few. In every case, the function of the protein depends on its shape. The shape of a protein, in turn, is the result of four levels of structure: primary, secondary, tertiary, and quaternary.

Primary structure results from the sequence of amino acids that make up the protein chain.

Secondary structure results from the hydrogen bonding within the molecule. The helical nature of many proteins is the result of hydrogen bonding.

Tertiary structure is the intricate, three-dimensional shape or **conformation** of a protein and *most directly determines the way it functions and its specificity.* Enzymes **denature** (lose their natural shape) in high temperatures or adverse pH. When a protein/enzyme denatures, it cannot function because its tertiary structure has been altered beyond repair.

Quaternary structure refers to proteins that consist of more than one polypeptide chain. Hemoglobin, for example, exhibits quaternary structure because it consists of four chains; see Figure 1.10.

REMEMBER
Tertiary structure is directly responsible for the shape of a protein and how it functions.

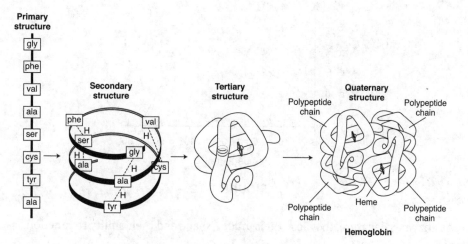

FIGURE 1.10 Protein structure

Enzymes

- Enzymes are large proteins.
- Enzymes serve to speed up reactions by lowering the **energy of activation** (E_a), which is the amount of energy needed to begin a reaction.
- The chemical that an enzyme works on is called a **substrate**.
- Enzymes are specific. In Figure 1.11, only substrate *A* will bind to the enzyme.
- The **induced-fit model** describes how enzymes work. As the substrate enters the active site, it induces the enzyme to alter its shape slightly so the substrate fits better. (The old "lock and key" model was abandoned because it implied that the enzyme never changes.)

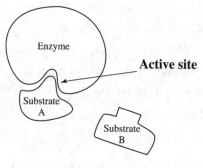

INTERESTING FACT

Some people cannot digest dairy products because they lack the enzyme lactase.

FIGURE 1.11

- Enzymes are not degraded during a reaction and are reused.
- Enzymes are named after their substrate, and the name ends in the suffix "ase." For example, sucrase is the name of the enzyme that hydrolyzes sucrose, and lactase is the name of the enzyme that hydrolyzes lactose.
- Enzymes function with the assistance from **cofactors** (minerals) or **coenzymes** (vitamins).
- The efficiency of the enzyme is affected by temperature and pH. Average human body temperature is 37°C, near optimal for human enzymes. If body temperature rises above 40°C, the enzymes will stop functioning, as shown in Figure 1.12.

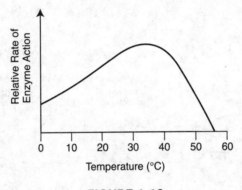

FIGURE 1.12

- As enzymes denature, they lose their unique shape and their ability to function.
- Gastric enzymes become active at low pH, when mixed with stomach acid. In contrast, intestinal amylase works best in an alkaline environment; see Figure 1.13.

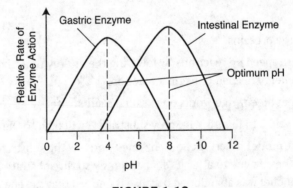

FIGURE 1.13

Prions—Proteins That Cause Disease

Prions are infectious proteins that cause several brain diseases, including mad cow disease. A prion is a misfolded version of a protein normally found in the brains of mammals. If a prion gets into a normal brain, it causes all the normal proteins to misfold in the same way.

Nucleic Acids

The nucleic acids are deoxyribonucleic acid (DNA) and ribonucleic acid (RNA). They carry hereditary information. (See Figure 1.14.)

- They are **polymers** (chains of repeating units) of **nucleotides**.

- A single nucleotide consists of a phosphate, a 5-carbon sugar (either deoxyribose or ribose), and a nitrogenous base.

- In DNA, the nitrogen bases are adenine, cytosine, guanine, and thymine.

- In RNA, the bases are adenine, cytosine, guanine, and uracil, in place of thymine.

- Adenine and guanine are purines.

- Cytosine, thymine, and uracil are pyrimidines.

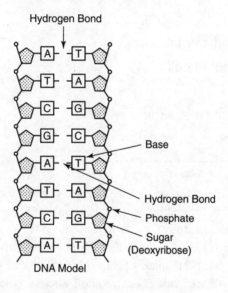

FIGURE 1.14 DNA molecule

The Cell

Learning Objectives

In this chapter, you will learn:

- Cell theory
- Structure of plant and animal cells
- Transport into and out of cells
- Life functions
- The metric system and measuring
- Tools and techniques to study cells

In seventeenth-century Holland, Anton van Leeuwenhoek, using lenses he crafted, was the first person to observe and document living cells. In about 1665, Robert Hooke developed a microscope that enabled him to study cork tissue. He coined the term "cell." The German botanist Matthias Schleiden (1838), studying living tissue using these newly developed microscopes, concluded that all plants are made of cells. About the same time, Theodor Schwann (1839) stated that all animals are made of cells. Rudolf Virchow (1855), a pathologist studying cell reproduction, summarized his years of research by stating, "Where a cell exists, there must be a pre-existing cell." The work and discoveries of all these scientists are responsible for one of the fundamental theories of biology, the cell theory.

Cell Theory

Modern cell theory states:

- All living things are composed of cells
- Cells are the basic units of all organisms
- All cells arise from preexisting cells

Most plant and animal cells have diameters between 10 and 100 micrometers (μm). Some, though, like red blood cells, are very small, with a diameter of 8 μm. All cells are enclosed by a membrane that regulates the passage of materials between a cell and its surroundings. They also contain nucleic acid, which directs the cell's activities and controls inheritance.

Cells are divided into two varieties: prokaryotes and eukaryotes. **Prokaryotes** have no nucleus or other internal membranes. All bacteria are prokaryotes. **Eukaryotes** have a nucleus and are more complex cells. They make up every form of life other than bacteria. Human cells are eukaryotic cells. Figure 2.1 shows a typical prokaryotic bacterial cell that lacks a nucleus and all membrane-bound internal structures. Figure 2.3 and Figure 2.4 show typical plant and animal cells, respectively. Table 2.1 compares prokaryotic and eukaryotic cells.

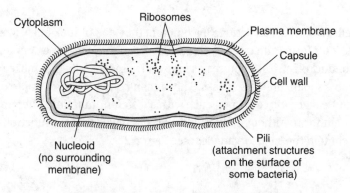

FIGURE 2.1 Typical prokaryotic cell

TABLE 2.1 Comparison of Prokaryotes and Eukaryotes

Prokaryotes	Eukaryotes
No membrane-bound organelles such as a nucleus	Contain distinct organelles surrounded by membranes, such as a nucleus and mitochondria
Contains a single, circular chromosome; may also contain plasmids (extrachromosomal DNA)	Chromosomes are linear; human body cells contain 46 chromosomes in each nucleus
Can contain plasmids	Does not contain plasmids
Ribosomes are smaller	Ribosomes are larger
Respiration can be either aerobic or anaerobic	Respiration is mostly aerobic
Cytoskeletal elements, such as microfilaments, are absent	Cytoskeletal elements, like microfilaments and microtubules, are present
Most are unicellular	Some, like euglena and paramecia, are single celled; many are multicellular with specialized cell types, such as muscle, blood, and skin cells
Very small: 1–10 μm	Larger: 10–100 μm
Most have tough external cell walls	Most (except plant cells and protists) are surrounded by only a cell membrane

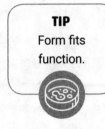

TIP

Form fits function.

Scientific studies using radioactive dating indicate that Earth is about 4.6 billion years old. All organisms on Earth are believed to have descended from a common ancestral prokaryotic cell about 3.5 billion years ago. According to the **theory of endosymbiosis**, eukaryotic cells containing organelles like mitochondria and chloroplasts evolved when free-living prokaryotes took up permanent residence inside other larger prokaryotic cells, about 2 billion years ago. This was the origin of a complex eukaryotic cell with internal membranes that compartmentalized the cell, making it very efficient and leading to the evolution of all multicelled organisms.

Although all cells have many structures in common, they do not look alike. Many, in fact, are quite unique. A cell's form is dictated by its function. A nerve cell, for example, whose purpose is to conduct electrical impulses, is long and spindly. Cells that make up a tough peach pit resemble square building blocks. The human body is made of approximately two hundred different eukaryotic cell types, each with a different function. Therefore, each cell type has a different form. Although different cell types have different functions and appearances, they all contain the same organelles; see Figure 2.2.

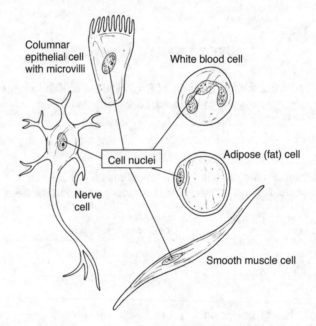

FIGURE 2.2 Five different eukaryotic cell types

Structures of Plant and Animal Cells

Plant and animal cells have many cellular organelles in common, including ribosomes and mitochondria. However, they also have organelles unique to the cell type, such as cell walls in plant cells and centrioles in animal cells. Figure 2.3 and Figure 2.4 show typical plant and animal cells, respectively.

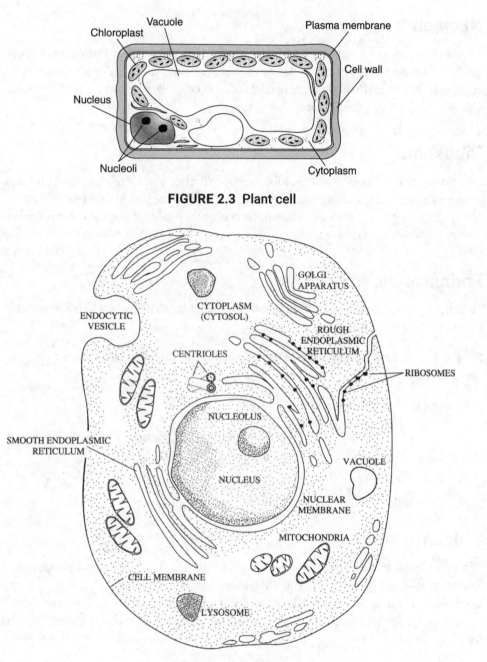

FIGURE 2.3 Plant cell

FIGURE 2.4 Animal cell

Nucleus

The nucleus contains chromosomes made of DNA that is wrapped with special proteins called histones into a **chromatin network**. Chromosomes contain genes, bits of DNA that code for polypeptides. The nucleus is surrounded by a selectively permeable double membrane or envelope that contains pores and allows for the transport of large molecules, such as RNA, out of the nucleus and into the cytoplasm.

Nucleolus

The **nucleolus** is a prominent region inside the nucleus of a cell that is not dividing. Components of **ribosomes** are synthesized here. Nucleoli are not membrane-bound structures but are tangles of chromatin and unfinished bits of ribosomes. One or two nucleoli are commonly visible in a nondividing cell.

Ribosome

This is the site of protein synthesis. Ribosomes are particles made of ribosomal RNA and protein. They are either suspended freely in the cytoplasm or bound to the endoplasmic reticulum. A single cell, such as a human liver cell, that produces large amounts of protein contains millions of ribosomes.

Endoplasmic Reticulum

The **endoplasmic reticulum** (ER) is a system of membrane channels that traverse the cytoplasm. There are two varieties.

Rough ER is studded with ribosomes. Therefore, it is the site of protein synthesis as well as transport throughout the cytoplasm.

Smooth ER has many functions:

1. Synthesizes steroid hormones and other lipids
2. Connects rough ER to the Golgi apparatus
3. Detoxifies the cell
4. Carbohydrate (glycogen) metabolism

Golgi Apparatus

The **Golgi apparatus** lies near the nucleus and consists of flattened sacs of membranes stacked next to each other (like a stack of pancakes) and surrounded by vesicles. They modify, store, and package substances produced in the rough endoplasmic reticulum. The Golgi apparatus also secretes these substances into other parts of the cell and onto the cell surface for export to other cells.

Lysosome

A lysosome is a sac of hydrolytic (digestive) enzymes enclosed by a single membrane. It is the principal site of **intracellular digestion**. With the help of the lysosome, the cell continually renews itself by breaking down and recycling cell parts. Programmed cell death (**apoptosis**) is critical to the embryonic development of multicellular organisms and is carried out by a cell's own hydrolytic enzymes. Plant cells do not usually have lysosomes.

Mitochondrion

The **mitochondrion** (plural, mitochondria) is the site of cellular respiration. All cells have many mitochondria. A very active cell could have about 2,500 of them. Mitochondria consist of an outer double membrane and folded inner membranes called **cristae**. Enzymes that are important to cellular respiration are embedded in the cristae membrane. Mitochondria contain their own DNA and can self-replicate. See Figure 2.5. (Remember, they were free-living prokaryotes several billion years ago.)

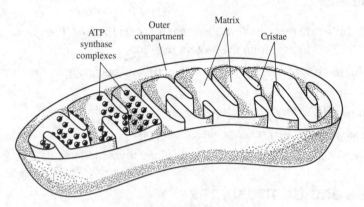

FIGURE 2.5 Mitochondrion

Vacuole

Vacuoles are single, membrane-bound structures that store substances for the cell. Fresh-water protista, like paramecia and amoebas, have **contractile vacuoles** that pump excess water out of the cell. Plant cells and some specialized human (fat or adipose) cells have large central vacuoles for storage.

Vesicle

Vesicles are tiny vacuoles. They are found in many places in cells, including the axon of a neuron, where they release neurotransmitters into a synapse.

Plastids

Plastids have a double membrane and are found in only plants and algae. There are three types.

1. **Chloroplasts** are green because they contain chlorophyll. They are the sites of photosynthesis. In addition to a double outer membrane, they have an inner one that forms a series of structures called **grana**. The grana lie in the stroma. Chloroplasts, like mitochondria, contain their own nuclear material and can self-replicate. (Remember, they were free-living prokaryotes several billion years ago.)

2. **Leucoplasts** are colorless and store starch. They are found in roots, like turnips, or in tubers, like potatoes.

3. **Chromoplasts** store carotenoid pigments and are responsible for the red-orange-yellow coloring of carrots, tomatoes, daffodils, and many other plants. These bright pigments in petals attract pollinating insects to flowers.

Cytoskeleton

The cytoskeleton of a cell is a complex network of protein filaments that extends throughout the cytoplasm and gives the cell its shape and enables it to move. The cytoskeleton includes two types of structures.

1. **Microtubules** are thick, hollow tubes that make up the cilia, flagella, and spindle fibers. They are formed from the protein tubulin.

2. **Microfilaments** are made of the protein **actin** and help support the shape of the cell. They enable

   - Animal cells to form a **cleavage furrow** during cell division
   - Amoebas to move by sending out pseudopods
   - Skeletal muscles to contract by sliding along **myosin** filaments

Centrioles and Centrosomes

> **REMEMBER**
> Centrioles =
> 9 triplets

Centrioles and centrosomes lie outside the nuclear membrane and organize the spindle fibers required for cell division. Only animal cells have centrioles and centrosomes. Plant cells have microtubule-organizing regions instead. Two **centrioles**, at right angles to each other, make up one **centrosome**. Centrioles and spindle fibers have the same structure. As shown in Figure 2.6, they consist of 9 triplets of **microtubules** arranged in a circle.

> **REMEMBER**
> Cilia and
> flagella = 9 + 2

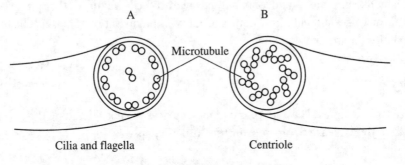

FIGURE 2.6

Cilia and Flagella

Cilia and flagella have the same internal structure; both are made of microtubules. The only structural difference is in the length: cilia are short, and flagella are long. Figure 2.6 shows that both consist of 9 pairs of microtubules organized around 2 singlet microtubules.

Cell Wall

The cell wall is one structure not found in animal cells. Cell walls of fungi consist of **chitin**, while plants and algae have cell walls made of **cellulose**. In plant cells, the primary cell wall is located immediately outside the plasma membrane. Some cells produce a second cell wall underneath the primary cell wall called the **secondary cell wall**. When a plant cell divides, a thin gluey layer is formed between the two cell walls, which becomes the **middle lamella** and which keeps the two daughter cells attached.

Cytoplasm and Cytosol

The entire region between the nucleus and plasma membrane is called **cytoplasm**. **Cytosol** refers to the semiliquid portion of the cytoplasm. In eukaryotic cells, organelles are suspended in the cytosol and get carried around the cell as the cytoplasm cycles around the cell, a process called **cyclosis**.

Cell or Plasma Membrane

The cell or plasma membrane is a selectively permeable membrane that controls what enters and leaves the cell. It is described as a **fluid mosaic** because it is made of many small particles that are able to move around in order to control what enters and leaves the cell. The plasma membrane consists of a **phospholipid bilayer** with proteins dispersed throughout. Molecules of **cholesterol** are embedded within the membrane, making it less fluid and more stable. The external surface of the plasma membrane has carbohydrate chains attached to it that are important for cell-to-cell recognition, as can be seen in Figure 2.7.

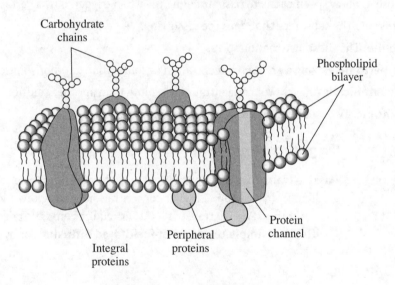

FIGURE 2.7 Plasma membrane

An average cell membrane consists of about 60 percent protein. These proteins provide a wide range of functions for the cell. Some membrane proteins, like **ATP synthase**, act as an enzyme. Some, like those involved in the **sodium-potassium pump**, transport ions into and out of cells.

Table 2.2 shows the differences between plant and animal cells.

TABLE 2.2 Differences Between Animal and Plant Cells

Animal Cells	Plant Cells
Centrioles and centrosomes	No centrioles or centrosomes
No chloroplasts and other plastids	Chloroplasts and other plastids
Most have small vacuoles (specialized fat storage cells are an exception)	Large central vacuoles
Plasma membrane only	Cell wall in addition to plasma membrane
Lysosomes	No lysosomes

Transport Into and Out of the Cell

Here is some important vocabulary for a discussion about transport.

1. **Selectively permeable.** A characteristic of a living membrane. The substances that pass through a selectively permeable membrane change with the needs of a cell. For example, the axon membrane of a nerve cell consists of gated channels that open and close to allow specific ions to pass through only when triggered by a certain stimulus.

2. **Solvent.** The substance that does the dissolving.

3. **Solute.** The substance that dissolves.

4. **Hypertonic.** Having a greater concentration of solute than another solution.

5. **Hypotonic.** Having a lower concentration of solute than another solution.

6. **Isotonic.** Two solutions containing equal concentrations of solute.

Passive Transport

DON'T BE FOOLED

Facilitated diffusion does not require ATP/energy.

Passive transport is the movement of molecules down a concentration gradient from a region of higher concentration to a region of lower concentration. Passive transport NEVER requires energy. It occurs either by **diffusion** (**simple diffusion** or **facilitated diffusion**) or by **osmosis**.

Simple Diffusion

Simple diffusion is merely the movement of particles from a higher concentration to a lower concentration. The steeper the gradient, the faster the rate of diffusion. Earthworms "breathe" as oxygen from the air is absorbed by simple diffusion across their moist skin into capillaries directly beneath the skin. Humans obtain oxygen by simple diffusion across moist membranes in air sacs, called **alveoli**, in our lungs.

Facilitated Diffusion

Facilitated diffusion relies on special **protein** membrane **channels** to assist in transporting specific substances across a membrane. For example, the normal functioning of a neuron requires calcium ions to be transported by facilitated diffusion through calcium ion channels within the axon membrane. Figure 2.8 shows a protein channel.

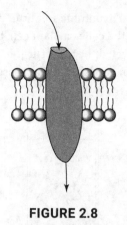

FIGURE 2.8

Osmosis

Osmosis is the diffusion of water across a membrane. Water flows down a gradient toward a region with high solute concentration. For example, in Figure 2.9, cell A has more solute than cell B. Therefore, cell A is **hypertonic** to cell B and cell B is **hypotonic** to cell A. Water will flow toward the region of higher concentration of solute, from cell B to cell A.

Diffusion of water

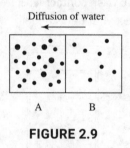

A   B

FIGURE 2.9

> **REMEMBER**
> Water diffuses toward the hypertonic area.

Figure 2.10 is a diagram of a cell in a hypertonic solution. Water will leave the cell, causing the cell to shrink. This cell shrinking is known as **plasmolysis**. In class, you may have carried out an experiment where you dropped a solution of 5 percent sodium chloride onto a living cell (such as elodea) while you watched under a microscope. Doing this caused the cell to shrink.

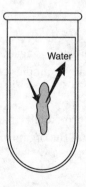

Cell in a hypertonic solution

FIGURE 2.10

Figure 2.11 is a diagram of a cell in a hypotonic solution. Water flows into the cell. This causes an animal cell to burst. However, if the cell is a plant cell, the cell wall will prevent the cell from bursting. Plant cells merely swell or become **turgid**. This turgid pressure is what keeps vegetables like celery or green peppers crisp. If a plant loses too much water (dehydrates), it loses **turgor pressure** and wilts.

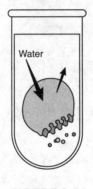

Cell in a hypotonic solution

FIGURE 2.11

Figure 2.12 shows a cell in an isotonic solution. Water diffuses in and out, but there is no net change in the cell. Solutions used to wash contact lenses or saline solutions used to wash out your eyes must be isotonic to protect your delicate body cells so as not to cause any damage.

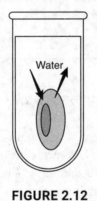

Cell in an isotonic solution

FIGURE 2.12

Active Transport

Active transport is the movement of molecules against a gradient, which requires energy, usually in the form of ATP. There are many examples of active transport in the biology course you studied.

The **contractile vacuole** in freshwater protista like paramecia and amoebas pumps out excess water that diffuses inward because the organisms live in an environment that is hypotonic.

Exocytosis

Exocytosis is the active release of molecules from a cell. A good example is found in the synapse of nerve cells. Vesicles containing a neurotransmitter such as acetylcholine (ACh) release their contents into the synapse in order to pass an impulse to another cell.

Endocytosis

Endocytosis is the process by which cells take in various molecules and particles by forming new vesicles made from the plasma membrane. There are three examples: **pinocytosis**, **phagocytosis**, and **receptor-mediated endocytosis**.

Pinocytosis

Pinocytosis, also called cell drinking, is the uptake of small, dissolved molecules. The plasma membrane invaginates around tiny particles and encloses them in a **vesicle**, as shown in Figure 2.13.

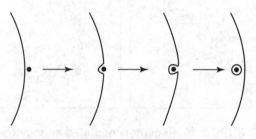

FIGURE 2.13 Pinocytosis

Phagocytosis

Phagocytosis is the engulfing of large particles or even small organisms by **pseudopods**. As shown in Figure 2.14, the cell membrane wraps around the particles and encloses them, forming a vacuole. This is the way human white blood cells engulf bacteria and also the way in which amoebas feed.

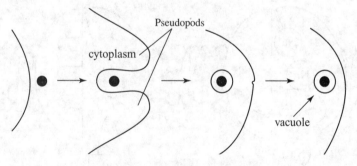

FIGURE 2.14 Phagocytosis

Receptor-mediated endocytosis

Receptor-mediated endocytosis enables a cell to take up large quantities of very specific substances. Extracellular substances bind to specific receptors on the cell membrane and are drawn into the cell into vesicles, as seen in Figure 2.15. This is the way in which body cells take up cholesterol from the blood.

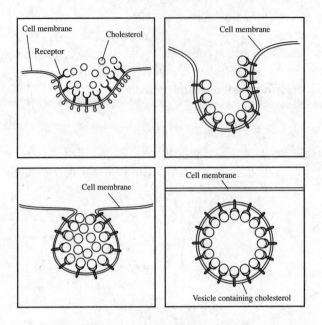

FIGURE 2.15 Receptor-mediated endocytosis

Another important example of how cells actively transport particles or ions across a membrane against a gradient is the **sodium-potassium pump.** The sodium-potassium pump in nerve cells, seen in Figure 2.16, carries sodium (Na^+) and potassium (K^+) across the axon membrane in opposite directions to return the nerve to its resting state after an impulse has passed.

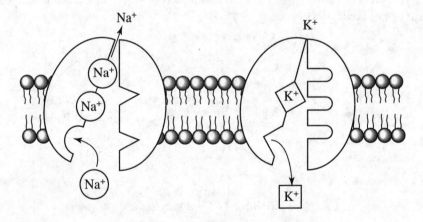

FIGURE 2.16 Sodium-potassium pump

Figure 2.17 shows an overview of passive and active transport.

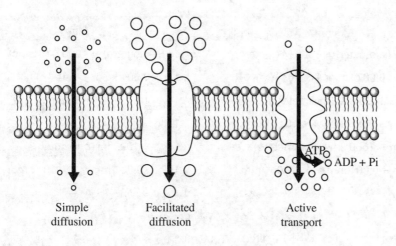

FIGURE 2.17

The Life Functions

A characteristic of all cells is that they carry out certain life processes. They include:

1. **Ingestion.** Intake of nutrients
2. **Digestion.** Enzymatic breakdown (hydrolysis) of food so it is small enough to be assimilated by the body
3. **Respiration.** Metabolic processes that produce energy (adenosine triphosphate or ATP) for all the life processes
4. **Transport.** Distribution of molecules from one part of a cell to another or from one cell to another
5. **Regulation.** Ability to maintain internal stability (homeostasis)
6. **Synthesis.** Combining of small molecules or substances into larger, more complex ones
7. **Excretion.** Removal of metabolic wastes
8. **Egestion.** Removal of undigested waste
9. **Reproduction.** Ability to generate offspring
10. **Irritability.** Ability to respond to stimuli
11. **Locomotion.** Moving from place to place (animal cells only)
12. **Metabolism.** Sum total of all the life functions

The Metric System and Measuring

Length	Mass
1 km (kilometer) = 1000 M (meters)	1 kg (kilogram) = 1000 g (grams)
1 M = 0.001 km or 1×10^{-3} km	1 g = 0.001 kg or 1×10^{-3} kg
1 M = 1000 mm (millimeters)	1 g = 1000 mg (milligrams)
1 mm = 0.001 M = 1×10^{-3} M	1 mg = 0.001 g or 1×10^{-3} g
1 mm = 1000 µm (micrometers)	1 mg = 1000 µg (micrograms)
1 µm = 0.001 mm or 1×10^{-3} mm	1 µg = 0.001 mg or 1×10^{-3} mg

Volume
1 L (liter) = 1000 mL
1 mL = 0.001 L or 1×10^{-3} L

Figure 2.18 is an image of a metric ruler showing the relationship of centimeters (cm) and millimeters (mm).

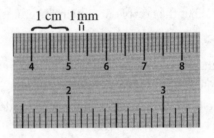

FIGURE 2.18

Measuring with a Ruler

Figure 2.19 is an image of a common ruler with inches on one edge and metric on the other. The metric edge is marked with *cm* for centimeters (1 M = 100 cm; 1 cm = 10 mm).

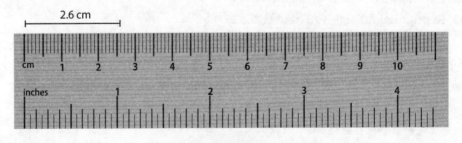

FIGURE 2.19

The line shown is 2.6 cm or 26.0 mm long.

Accurate and Precise

Saying that the line is 26.0 mm in length means that the line is *exactly* 26.0 mm long. Writing the value *without* the zero and decimal, simply as 26, means that the line is only *approximately* 26 mm long. It could actually be 25.9 or 26.1 mm long. In science, we try to be both *accurate* (correct) and *precise* (exact). To a scientist, these two terms mean very different things. For example, if you stated that you measured that same line and found it to be 25.55 mm long, your answer would be *precise* (to the hundredths place), but it would not be *accurate*. You would, in fact, be wrong.

Measuring with a Graduated Cylinder

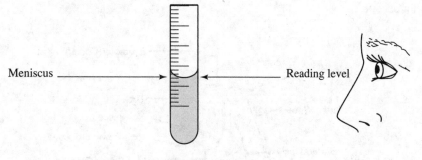

Meniscus ⟶ ⟵ Reading level

FIGURE 2.20

Use a *graduated cylinder* to measure liquid; and use the *meniscus* (bottom of the curve for most liquids) at eye level to take your measurement. (See Figure 2.20.)

Measure the amount of liquid in Figure 2.21 below: _____

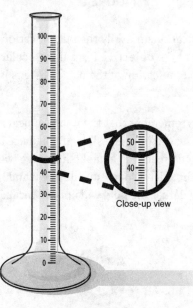

Close-up view

FIGURE 2.21

The correct answer (both accurate *and* precise) is 44.0 mL.

Tools and Techniques to Study Cells

There are many tools and techniques to study cells. But, the main tool for studying cell structure (cytology) is the **compound microscope**. Besides the ability to magnify an image, another important characteristic of a good microscope is the measure of image clarity, known as **resolution**. The finest microscopes have both high magnification and excellent resolution. A toy microscope, which may enlarge an image 400×, has little resolving power, so the images are blurred.

Anton van Leeuwenhoek developed the first microscope in the seventeenth century. Today, compound light microscopes have been refined. Even the microscope you may have used in high school has fine resolution and good magnification. Figure 2.22 is a picture of a compound microscope.

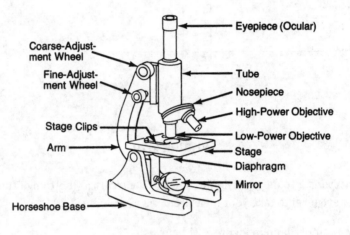

FIGURE 2.22

To determine the total magnification, multiply the magnification of the **ocular lens** or eyepiece by the magnification of the **objective lens**. If the ocular has a magnification of 10×, which is customary, and the magnification of the objective lens is 40×, the total magnification is 400×.

When you use the microscope, remember that the image appears upside down and backward from the actual specimen you placed onto the slide. If you place a letter "e" onto the slide so that it appears the way you would read it in a book, the image will appear like it does in Figure 2.23 when you look in the microscope—upside down and backward. Also, the higher the magnification you use, the darker the field will appear because you are viewing a much smaller area.

FIGURE 2.23

Today, the field of microscopy is very sophisticated. There are many different types of microscopes fashioned for different purposes.

1. Phase-contrast microscope
2. Transmission electron microscope
3. Scanning electron microscope

A **phase-contrast microscope** is a light microscope that enhances contrast. It is useful in examining living, *unstained cells*.

Electron microscopes use a beam of electrons, instead of a beam of light, to produce superior resolving power as well as magnification over $100,000\times$.

The **transmission electron microscope** (TEM) is useful for studying the interior of cells. The source of electrons is a tungsten filament within a vacuum column.

Although the TEM is very useful, there are some drawbacks:

- The tissue is no longer alive after processing.
- Preparation of specimens is elaborate. Tissue must be fixed, dehydrated, and sectioned on a special machine, a process that requires many hours and much expertise.
- The TEM is a delicate machine and requires special engineers to maintain it.
- Specimens must be sliced so thin that only a small portion of a tissue sample can be studied at one time.
- The machine costs hundreds of thousands of dollars.

DON'T FORGET

Specimens observed under the electron microscope are not alive.

The **scanning electron microscope** (SEM) is useful for studying the surface of cells. The resulting images have a three-dimensional appearance. Once again, specimens are examined only after an elaborate process that kills the tissue.

Other Tools for Studying Cells

Another important tool used in the study of tissue is the **ultracentrifuge**. It enables scientists to isolate specific components of cells in *large quantities* by **cell fractionation**. By using this technique, hundreds of organelles, such as mitochondria, can be studied under an electron microscope or analyzed biochemically.

First, tissue is mashed in a blender. The resulting liquid, called homogenate, is spun at high speed in an ultracentrifuge and separated into layers based on differences in density. For example, if tissue is spun at high speed in a centrifuge tube, nuclei are forced to the bottom *first*, followed by mitochondria and then ribosomes with clear liquid above the organelles. This is shown in Figure 2.24.

Freeze fracture, also called freeze-etching, is a complex technique used to study details of membrane structure under an electron microscope. After preparation, only a cast of the original tissue is available to examine.

Tissue culture is a technique used to study the properties of specific cells in vitro (in the laboratory). Living cells are seeded onto a sterile culture medium to which a variety of nutrients and growth-stimulating factors have been added. Different cells require different growth media. Cell lines can be grown in culture for years provided great care is taken with them. While the cells are growing, they can be examined unstained under a phase-contrast light microscope.

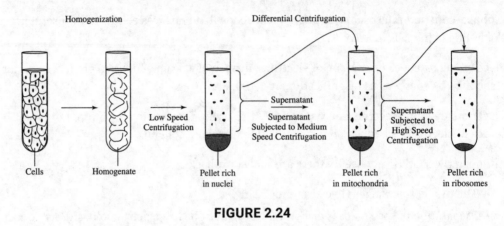

FIGURE 2.24

Cell Division— Mitosis and Meiosis

Learning Objectives

In this chapter, you will learn:
- Cell cycle
- Mitosis
- Meiosis

There are two types of cell division, mitosis and meiosis. **Mitosis** functions in the growth and repair of body cells. It produces two genetically identical daughter cells with the same chromosome number as the parent cell. Each daughter cell is diploid ($2n$), just like its parent cell.

Meiosis occurs only in sexually reproducing organisms. It produces gametes (sperm and ova) with half the chromosome number of the parent cell. Each resulting cell is **haploid** (n).

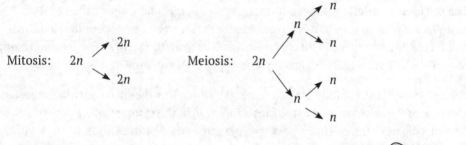

Any discussion about cell division must first include the structure of the chromosome. A chromosome consists of a highly coiled and condensed strand of DNA. A replicated chromosome consists of two **sister chromatids**, where one is an exact copy of the other. The **centromere** is a specialized region that holds the two sister chromatids together. Spindle fibers connect the centromere to the centrosome during cell division. Figure 3.1 shows a replicated chromosome.

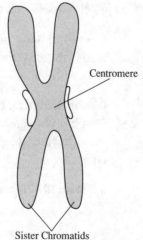

FIGURE 3.1 A replicated chromosome

33

Figure 3.2 is a cell containing four replicated chromosomes. Each chromosome contains two identical chromatids connected at a centromere.

FIGURE 3.2

The Cell Cycle

Living and dividing cells pass through a regular sequence of growth and division called the **cell cycle**. The rate and timing of the phases of cell division in an organism are crucial to normal growth, development, and maintenance.

The frequency of cell division varies with the cell type. For example, skin cells and cells that line the digestive tract divide constantly and continue to do so throughout one's life. Some specialized cells like nerve and muscle cells permanently lose the ability to divide and remain in the phase known as G_0 (pronounced G-zero). **Stem cells** from human embryos retain the ability to divide indefinitely and to differentiate into any cell type. Scientists are trying to coax stem cells to differentiate into nerve tissue in order to treat spinal cord injuries and Parkinson's disease.

The cell cycle is strictly regulated by many proteins, including some called **cyclins**. Some proteins are **internal regulators**, making sure that the cell does not undergo mitosis until certain conditions are met. Other proteins such as **growth factors** are **external regulators** of the cell cycle: speeding it up, slowing it down, or even stopping it.

The cell cycle consists of five stages: G_1, **S**, G_2 (which together make up interphase), **mitosis** (division of the nucleus), and **cytokinesis** (division of the cytoplasm). To move from one phase to the next, the cell must pass through three **checkpoints**. At each checkpoint, specialized proteins assess that the cell is ready to proceed to the next phase. See Figure 3.3.

G_1 or **First Gap Phase:** The newly divided cell enters this phase right after completing mitosis. During G_1, the cell increases in size and prepares to replicate its DNA.

G_1 **Checkpoint:** If the cell is healthy and contains adequate resources and if the DNA is undamaged, growth factors stimulate the cell to proceed to DNA synthesis during the S phase. Otherwise, the cell either dies or enters a resting state, referred to as G_0 phase.

S Phase: DNA is synthesized through the process of replication.

S Checkpoint: DNA synthesis is continuously monitored for any replication errors. If the synthesis progresses without errors, growth signals stimulate the cell to proceed to the next phase, G_2.

G_2 Phase or **Second Gap Phase:** The cell has to produce organelles and other cellular components that will populate two new functioning daughter cells.

G_2 Checkpoint: All chromosomes must be fully replicated and contain no other types of damage. Only then can the cell enter mitosis, or M (mitosis) phase, and divide.

Cell division is a strictly regulated process that mostly works without errors. When errors do occur, they can be catastrophic and cause the development of **cancer**.

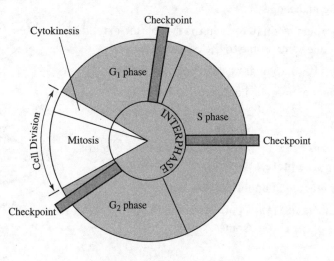

FIGURE 3.3 The cell cycle

Interphase

More than 90 percent of the life of a cell is spent in interphase. Most cells you have observed under a microscope are in this phase, including that shown in Figure 3.4.

During interphase, chromosomes replicate in preparation for cell division. One or more nucleoli become visible within the nucleus, and the nuclear membrane remains intact.

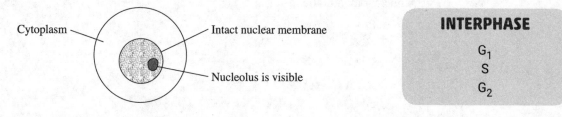

FIGURE 3.4 Interphase

INTERPHASE

G_1

S

G_2

REMEMBER THE STAGES OF MITOSIS

Prophase
Metaphase
Anaphase
Telophase

Mitosis

Mitosis is the actual division of the nucleus. As seen in Figure 3.5, it is arbitrarily divided into four phases: **prophase**, **metaphase**, **anaphase**, and **telophase**.

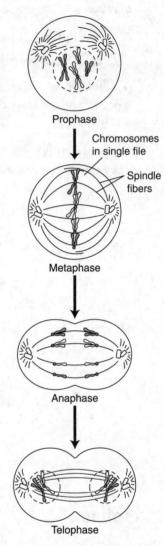

Prophase

1. Strands of chromosomes begin to condense and become visible.

2. The nucleoli disappear.

3. **Spindle fibers** begin to form in the cytoplasm, extending from one centrosome to the other.

4. Centrioles begin to migrate to the poles.

5. The nuclear membrane begins to disintegrate.

Metaphase

1. Chromosomes line up *single file* located on the equator or **metaphase plate**.

2. Centrosomes are at opposite poles of the cell.

3. Spindle fibers run from centrosomes to the centromeres of the chromosomes.

Anaphase

The centromeres of each chromosome separate, and spindle fibers begin to pull the sister chromosomes apart.

Telophase

1. Chromosomes cluster at opposite ends of the cell, and the nuclear membrane re-forms.

2. Supercoiled chromosomes begin to unravel and to return to their pre-cell division condition as long, threadlike strands.

3. The nuclear membrane re-forms.

FIGURE 3.5 Mitosis

Cytokinesis

Cytokinesis is division of the cytoplasm. In animal cells, a **cleavage furrow** forms down the middle of the cell as the cytoplasm pinches inward and the two daughter cells separate from each other, as shown in Figure 3.6 (left). In plant cells, a **cell plate** forms down the middle of the cell. Daughter cells do not separate from each other. Instead, a sticky **middle lamella** cements adjacent cells together, as shown in Figure 3.6 (right).

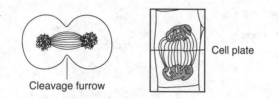

Cell plate

Cleavage furrow

FIGURE 3.6 Cytokinesis

Meiosis

Meiosis is a form of cell division in which cells having the diploid chromosome number (2*n*) produce gametes (sex cells) with the haploid (monoploid) chromosome number (*n*). Meiosis occurs in two stages, meiosis I and meiosis II.

Meiosis I

1. This stage is also called **reduction division**.
2. **Synapsis** and **crossing-over** occur. During synapsis, chromosomes pair up precisely with their homologues so that crossing-over can occur. Crossing-over is the process in which homologous chromatids exchange genetic material. Crossing-over is important because it ensures greater variety in the gametes.
3. *Homologous chromosomes separate.* Failure to separate is **nondisjunction**.
4. Next, chromosomes line up randomly on the equatorial plate and separate independently. This means that how one pair of chromosomes lines up and separates has no effect on how any other pair of chromosomes lines up and separates.
5. Each resulting gamete is genetically unique.

Meiosis II

1. This stage is similar to mitosis but does not have a special name.
2. *Sister chromatids separate.*
3. This division maintains the haploid number of chromosomes.
4. This phase completes the goal of meiosis—producing four genetically unique cells from one original mother cell.

> **REMEMBER**
> Mitosis is for growth and repair and meiosis is for the production of gametes only.

Figure 3.7 shows meiotic cell division with one cross-over event during prophase I. (Not all stages are shown.)

Pay particular attention to the shaded cells, metaphase I and metaphase II. In metaphase I of meiosis I, chromosomes are lined up **double file**. In metaphase II of meiosis II, chromosomes are lined up **single file**.

THIS IS TRICKY

Meiosis I
Homologous pairs separate

Meiosis II
Sister chromatids separate

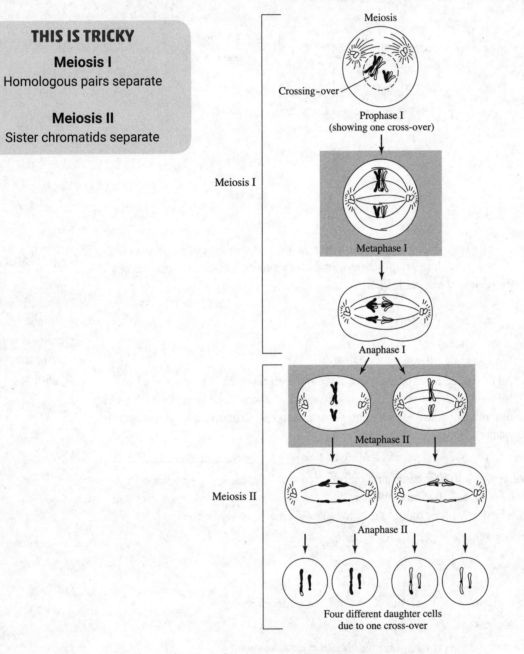

FIGURE 3.7 Meiosis

Cell Respiration

Learning Objectives

In this chapter, you will learn:

- ATP—adenosine triphosphate
- Structure of the mitochondrion
- Aerobic respiration—when oxygen is present
- NADH and $FADH_2$
- Anaerobic respiration—when oxygen is not present

Cell respiration is a series of oxidative reactions by which cells gradually release energy stored in glucose and *transfer it* to molecules of **ATP** (adenosine triphosphate). Energy stored in ATP is immediately available for cellular activities such as contracting muscles, passing an impulse along a nerve, or pumping ions by active transport.

You should know the equation for the complete aerobic respiration of one molecule of glucose. Here it is:

$$C_6H_{12}O_6 + 6O_2 \quad \rightarrow \quad 6CO_2 + 6H_2O + \text{energy}$$

As you can see, glucose combines with oxygen to produce ATP, an energy source, plus the waste products carbon dioxide and water.

ATP—Adenosine Triphosphate

ATP is the special high-energy molecule that stores energy for immediate use in the cell. It consists of **adenosine** (the nucleotide adenine + ribose) plus **3 phosphates**. Figure 4.1 shows the structure of a molecule of ATP.

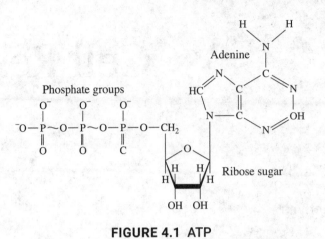

FIGURE 4.1 ATP

The removal of one phosphate group from ATP results in the formation of a more stable and lower energy molecule, ADP. Energy is released as ATP converts to ADP. Energy is absorbed to add a phosphate to ADP to produce ATP. Figure 4.2 shows this conversion.

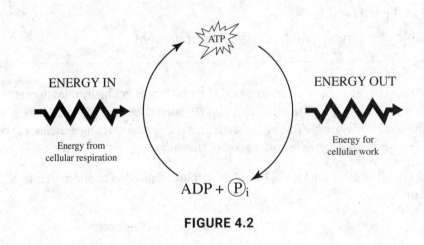

FIGURE 4.2

Structure of the Mitochondrion

The mitochondrion is enclosed by two membranes, an outer membrane and an inner **cristae** membrane that is folded. This inner membrane divides the mitochondrion into two internal compartments, the **outer compartment** and the **matrix**. The citric acid (Krebs) cycle takes place in the matrix; the electron transport chain takes place in the cristae membrane. Figure 4.3 is a diagram of the mitochondrion.

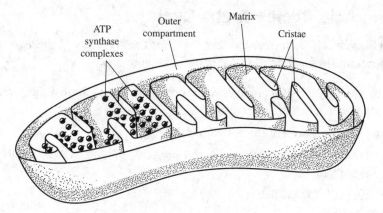

FIGURE 4.3 Mitochondrion

If oxygen is not present (anaerobic), glycolysis is followed by alcohol fermentation or lactic acid fermentation. If oxygen is present (aerobic), glycolysis is followed by the citric acid cycle, the electron transport chain, and chemiosmosis. A description of aerobic respiration follows.

Aerobic Respiration—When Oxygen Is Present

Aerobic cell respiration consists of two phases: an **anaerobic phase** and an **aerobic phase**. The anaerobic phase consists of **glycolysis**. When oxygen is present, glycolysis is followed by the aerobic phase that consists of the **citric acid cycle** and **chemiosmosis** or **oxidative phosphorylation**.

Glycolysis

Glycolysis is the anaerobic phase of aerobic respiration. One molecule of glucose breaks apart into two molecules of pyruvate. **Pyruvate**, or **pyruvic acid**, is essentially one-half a glucose molecule and is the *raw material for the next step in respiration*, the citric acid cycle.

- Glycolysis occurs in the cytoplasm of all cells.
- Glycolysis is a complex, multistep process, each step of which is controlled by a different enzyme.
- Two molecules of ATP supply the *energy of activation*, the energy needed to begin the reaction.
- Glycolysis releases 4 ATP molecules, resulting in a net gain of 2 ATP.
- 1 Glucose + 2 ATP → 2 Pyruvate + 4 ATP + 2 NADH (net gain 2 ATP).

REMEMBER

Glycolysis produces pyruvic acid and a small amount of ATP.

The Citric Acid Cycle—Krebs Cycle

The **citric acid cycle**, also known as the Krebs cycle, is the first stage of the *aerobic phase* of cellular respiration. The citric acid cycle is shown in Figure 4.4.

- Pyruvate (from glycolysis) combines with **coenzyme A** (a vitamin derivative) to form **acetyl-CoA**, which enters the citric acid cycle.

- This occurs in the matrix of the mitochondria.

- Each turn of the citric acid cycle produces 1 molecule of both ATP and $FADH_2$ plus 3 molecules of NADH.

- The by-product is CO_2, which is exhaled.

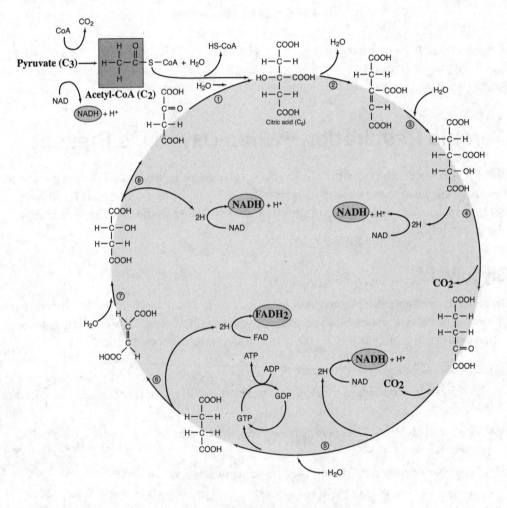

TIP
The citric acid cycle produces • a small amount of ATP • carbon dioxide • NADH and $FADH_2$

FIGURE 4.4 Citric acid cycle

NADH and FADH$_2$

NADH and **FADH$_2$** molecules carry out an important function in cell respiration.

- NADH and FADH$_2$ are **coenzymes** that shuttle protons and electrons from glycolysis and the citric acid cycle to the electron transport chain.

- NAD$^+$ is the oxidized form; NADH is the reduced form. FAD$^+$ is the oxidized form; FADH$_2$ is the reduced form.

- Ultimately, each NADH translates to the production of 3 ATP, while each FADH translates to the production of 2 ATP.

Electron Transport Chain, Chemiosmosis, and Oxidative Phosphorylation

FADH$_2$ and **NADH** transport highly energized electrons from the **citric acid cycle** to the **electron transport chains (ETCs)** within **cristae membranes** of mitochondria. The energy from these highly energized electrons *being pulled through* the ETC powers the pumping of protons across the **cristae membranes** from the **inner matrix** to the **outer compartment** *against a gradient*. The whole purpose of the ETC is to establish a **proton gradient**. Once formed, the gradient is maintained because protons cannot flow back through the cristae membrane except through special proton (H$^+$) channels called **ATP synthase channels**. See Figure 4.5.

Proton Gradient Represents Potential Energy

The proton gradient established by the electron transport chain in mitochondria represents enormous potential energy to power the phosphorylation of ADP to ATP (to make ATP). Similar to a hydroelectric plant, when water flows over a dam (pulled downward by gravity) it turns massive turbines that transfer kinetic energy (energy of movement) into electrical energy. In the outer compartment of mitochondria, when protons flow down a gradient through **ATP synthase channels**, their potential energy is transferred to kinetic energy and then to chemical bond energy stored in molecules of ATP. The production of ATP by this mechanism is called **chemiosmosis** or **oxidative phosphorylation**. Chemiosmosis and oxidative phosphorylation produce most of the ATP in aerobic cellular respiration.

Oxygen has a strong attraction for electrons and protons. It pulls the electrons through the ETC and also serves as the *final electron and proton acceptor* in the ETC. When oxygen combines with protons and electrons at the end of the ETC, water is formed as a waste product. This is the water vapor we constantly exhale.

Details of Cell Respiration

- The ETC is a collection of carrier molecules (including **cytochromes**) embedded in the **cristae membrane** of mitochondria.

- Every mitochondrion contains thousands of ETCs.

- The ETC carries electrons through a series of **redox reactions** as special molecules bind to and let go of electrons. In a redox reaction, one atom gains electrons (**reduction**) while one atom loses electrons (**oxidation**).

- **Water** is produced as a waste product as oxygen combines with protons and electrons that flow down the ETC. Here is the equation:

$$\frac{1}{2}O_2 + H_2 \rightarrow H_2O$$

- Protons cannot diffuse directly through the cristae membrane but can cross only through **ATP synthase channels**.

- Hypothetically, each proton carried by an NAD molecule to the ETC produces 3 ATP molecules, while each proton carried by an FAD molecule produces 2 ATP molecules.

REMEMBER

Chemiosmosis and phosphorylation produce the most ATP during cellular respiration.

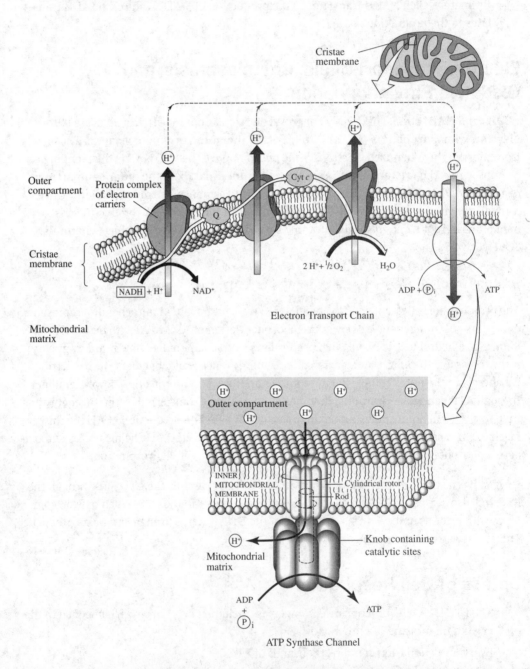

FIGURE 4.5 Electron transport chain, chemiosmosis, and oxidative phosphorylation

During respiration, most energy flows in this sequence:

Glucose $\rightarrow$ NAD and FAD $\rightarrow$ Electron transport chain (chemiosmosis) $\rightarrow$ ATP

Figure 4.6 provides an overview of aerobic respiration.

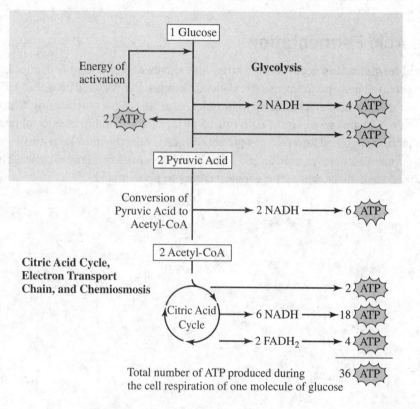

FIGURE 4.6 The complete breakdown of glucose during cell respiration

Anaerobic Respiration—When Oxygen Is Not Present

Anaerobic respiration or **fermentation** consists of the process known as **glycolysis** plus either **alcohol fermentation** or **lactic acid fermentation**. Anaerobic respiration originated billions of years ago when there was no free oxygen in Earth's atmosphere. Today it is the sole means by which anaerobic bacteria such as *Clostridium botulinum* (the bacterium that causes a form of food poisoning, botulism) release energy from food. Figure 4.7 shows the two types of fermentation, alcohol fermentation and lactic acid fermentation.

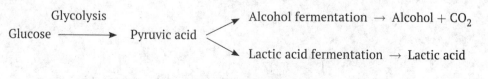

FIGURE 4.7

Alcohol Fermentation

Alcohol fermentation is the process by which certain cells convert pyruvic acid or pyruvate from glycolysis into ethyl alcohol and carbon dioxide in the absence of oxygen. The bread-baking industry depends on the ability of yeast to carry out fermentation and produce the carbon dioxide that causes bread to rise.

Lactic Acid Fermentation

Lactic acid fermentation occurs during strenuous exercise when the body cannot keep up with the increased demand for oxygen by skeletal muscles. Pyruvic acid produced by glycolysis converts to lactic acid and builds up in muscles, causing fatigue and burning. The expression "No pain, no gain" refers to the pain caused by lactic acid buildup in skeletal muscles. When an increase in blood flow restores proper oxygen levels, the muscle tissue reverts to the more efficient aerobic respiration and lactic acid is removed from the muscles. That lactic acid is carried to the liver, where it is converted back to pyruvic acid.

Photosynthesis

Photosynthesis is the process by which light energy is used to make glucose. Another way to describe it is to say that solar energy is converted into chemical energy because energy is stored in chemical bonds. Photosynthesis is carried out by all organisms in the Plantae kingdom as well as by algae in the Protista kingdom.

The general formula for photosynthesis is:

$$6CO_2 \quad + \quad 12H_2O \quad \rightarrow \quad C_6H_{12}O_6 \quad + \quad 6H_2O \quad + \quad 6O_2$$

carbon dixoide + water → sugar + water + oxygen

Overall, this process is a **reduction** reaction because the carbon in carbon dioxide is gaining electrons from the hydrogen in water. (The definition of reduction is the gain of electrons or gain of hydrogens.)

Structure of the Chloroplast

The chloroplast is an organelle enclosed by a double membrane. It contains **grana** (consisting of layers of membranes called **thylakoids**)—where the light-dependent reactions occur, and **stroma**—where the Calvin cycle (light-independent reactions) occurs. Figure 5.1 is a sketch of a chloroplast.

REMEMBER

Organisms that make their own food are called autotrophs.

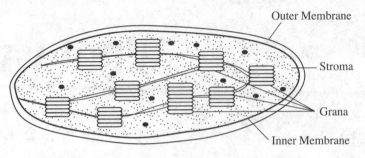

FIGURE 5.1 Chloroplast

Light and Photosynthetic Pigments

When light strikes an object, the light can be *reflected*, *transmitted*, or *absorbed*. Something that appears red reflects red light and absorbs all other colors of light. A green plant reflects green light and, therefore, cannot absorb green light as an energy source to make sugar. So, if you were to shine only green light onto a green plant, it would die for lack of energy to make its own food.

Photosynthetic pigments absorb light energy and use it to carry out photosynthesis. Substances that absorb light in the **visible spectrum** are called pigments. Different pigments absorb light of different wavelengths. However, only the green photosynthetic pigment, **chlorophyll *a*,** can participate directly in the light-dependent reactions. Other pigments, called **antennae** or **accessory pigments**, assist in photosynthesis by capturing and passing on photons of light to chlorophyll *a* and thus expand the range of light that can be used to produce sugar. These accessory pigments include chlorophyll *b*, carotenoids, and phycobilins. **Chlorophyll *b*** is green and absorbs all other wavelengths of light besides green. The **carotenoids** are yellow, orange, and red and are responsible for the color of carrots. The **phycobilins** are red and are found in red algae that live deep in the ocean where there is very little light.

Figure 5.2 is a graph showing the absorption spectrum for the photosynthetic pigments. Red and blue light are absorbed, while yellow and orange light are reflected. *Only light that is absorbed can be used to power the making of sugar.*

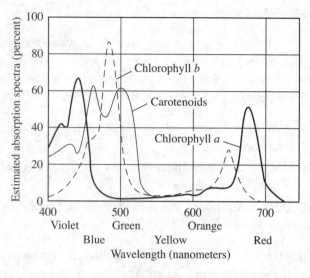

FIGURE 5.2

Light-Dependent and Light-Independent Reactions

There are two main processes of photosynthesis: the **light-dependent reactions**, or simply the light reactions, and the **light-independent reactions**. We used to call the light-independent reactions the "dark reactions." That name, however, is no longer used because it is misleading as *both reactions occur only when there is light*. In simplest terms, the function of the light-dependent reactions is to produce energy (ATP) and protons for the light-independent reactions. The function of the light-independent reactions is to make sugar (G3P).

Light-Dependent Reactions

The light-dependent reactions occur in the **grana** of chloroplasts within specialized membranes called **thylakoids**. Within these thylakoid membranes are hundreds of light-absorbing complexes called **photosystems**. Each photosystem consists of chlorophyll *a* and the accessory pigments, chlorophyll *b* and carotenoids. One of the main results of the light-dependent reactions is that large amounts of ATP are produced.

ATP is produced in chloroplasts the same way it is produced during cellular respiration in mitochondria—by electron transport chains and chemiosmosis. As the chlorophyll in the grana absorb light, electrons become energized and escape from chlorophyll molecules into **electron transport chains** (ETCs). The energy from these energized electrons pumps protons across the thylakoid membrane and creates a *proton gradient*. The potential energy stored in this proton gradient is converted into ATP as protons flow through an **ATP synthase** channel.

The following equation shows the flow of energy through the light-dependent reactions:

light → chlorophyll → energized electrons → ETC → proton gradient → ATP synthase → ATP

**LIGHT-DEPENDENT
REACTIONS**

1. Produce ATP
2. Provide protons to the light-
   independent reactions

Where does water come into the reactions?

Excited electrons escape from the chlorophyll molecules. How do they get replaced? In a nutshell, water is the source. Water breaks down, by a process called **photolysis**, into its components: electrons, protons, and oxygen atoms. Each has a role in making sugar.

1. **Electrons** replace those lost by chlorophyll in the light-dependent reactions.

2. **Protons** pass through the ATP synthase channels and get carried by **NADP** to the **stroma** for light-independent reactions.

3. **Oxygen** is released into the atmosphere as a waste product. This is the source of all our oxygen in the atmosphere.

Figure 5.3 is a sketch of the light-dependent reactions within the thylakoid membranes of chloroplasts. It shows the photosystems, electron transport chains, and the ATP synthase channels where ATP is produced.

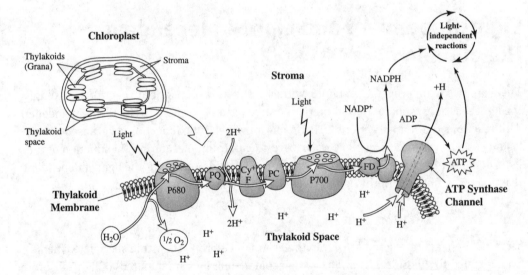

FIGURE 5.3 The light-dependent reactions

Light-Independent Reactions—Where Sugar is Made

TIP

Sugar is produced during the light-independent reactions.

The light-independent reactions occur in the **stroma**. Their function is to produce sugar or **G3P**, a 3-carbon sugar. In these reactions, CO_2, which the plant takes in through stomates in its leaves, combines with protons and electrons carried from the light reactions by NADP to produce sugar. Here is a simplified equation to show the process:

$$CO_2 + H^+ \text{ (protons)} + \text{electrons} \rightarrow G3P$$

The incorporation of carbon dioxide into a sugar molecule is called **carbon fixation**. It occurs during the cyclical process known as the **light-independent reactions**, as shown in Figure 5.4. Large amounts of ATP are required to keep the light-independent reactions running. All of the necessary ATP is produced during the light-dependent reactions.

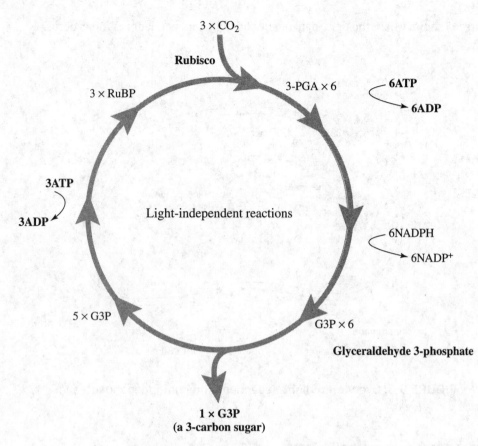

FIGURE 5.4 Light-independent reactions

The essence of the light-independent reactions:

- They take place in the stroma of chloroplasts.
- G3P (glyceraldehyde 3-phosphate), a 3-carbon sugar $C(H_2O)_n$, is produced.
- An all-important enzyme required for the light-independent reactions is **rubisco** (ribulose biphosphate).
- They use large amounts of ATP.

Table 5.1 compares cellular respiration and photosynthesis.

TABLE 5.1 Comparison of Cellular Respiration and Photosynthesis

Cellular Respiration	Photosynthesis
Occurs all the time	Occurs only in the light
Are oxidation reactions	Are reduction reactions
Occurs in mitochondria	Occurs in chloroplasts
Relies on ETC to produce a proton gradient	Relies on ETC to produce a proton gradient
Requires O_2 and releases CO_2	Requires CO_2 and releases O_2
NAD is the proton (and electron) carrier	NADP is the proton (and electron) carrier
ATP is produced by ATP synthase	ATP is produced by ATP synthase
Contains a cyclical process: citric acid cycle	Contains a cyclical process: light-independent reactions

Figure 5.5 shows where the light-dependent and light-independent reactions occur.

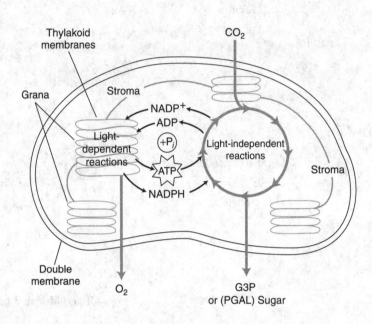

FIGURE 5.5 Overview of light-dependent and light-independent reactions

Structure of the Leaf

Figure 5.6 is a cross section of a typical leaf. Observe that the palisade layer consists of tightly packed cells that contain chloroplasts. This is where the majority of photosynthesis occurs. Like the **palisade layer**, the cells of the **spongy mesophyll** contain chloroplasts and also carry out photosynthesis. However, the cells in the spongy mesophyll are less tightly packed and are surrounded by **air spaces**, which allow for the exchange of oxygen, carbon dioxide, and water vapor. The epidermis layer is clear and does not carry out photosynthesis. It protects the delicate underlying cells and allows light to pass into the leaf. Above the epidermis is a waterproof layer of **cutin** that minimizes excessive water loss. **Guard cells** control the opening and closing of the **stomates**, which allow for the exchange of gases with a minimum of water loss.

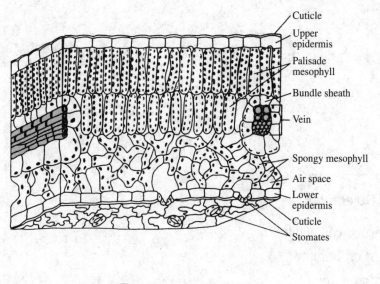

Cuticle
Upper epidermis
Palisade mesophyll
Bundle sheath
Vein
Spongy mesophyll
Air space
Lower epidermis
Cuticle
Stomates

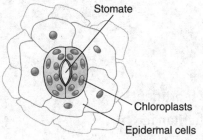

Stomate
Chloroplasts
Epidermal cells

FIGURE 5.6 The leaf

Classical Genetics

The father of modern genetics is Gregor Mendel, an Austrian monk. In the 1850s, he bred garden peas in order to study patterns of heredity. He collected data from hundreds of plants across many generations and applied statistical analysis to his carefully collected data. Mendel was successful and became famous because he brought mathematical laws of probability to the study of inheritance.

Mendel's work produced three laws: the **law of dominance**, the **law of segregation**, and the **law of independent assortment**.

Basics of Probability

Probability is the likelihood that a particular event will happen. It cannot predict whether a particular event will actually occur. However, if the sample is large enough, it can predict an average outcome.

If you flip a coin, there is a 50 percent chance that it will come up heads and a 50 percent chance that it will come up tails. The chance of either event happening is 1 out of 2 or $\frac{1}{2}$. If you flip a coin three times in a row, the probability of getting heads all three times is $\frac{1}{2} \times \frac{1}{2} \times \frac{1}{2}$ or $\frac{1}{8}$. You multiply, in this case, because each flip of the coin is a separate event. The chance that a couple will have a daughter is 50% or $\frac{1}{2}$. The chance that they will have 2 daughters is $\frac{1}{2} \times \frac{1}{2}$ or $\frac{1}{4}$. Again, you multiply because each birth is a separate event.

Understanding probability is important to the study of genetics because predicting outcomes is what Punnett squares are all about. What is the chance that two brown-eyed people can give birth to a child with blue eyes? That is probability.

Law of Dominance

Mendel's first law is the **law of dominance**. It states that when two organisms, homozygous (pure) for two opposing traits, are crossed, the offspring will be hybrid (carry two different **alleles**) but will exhibit only the dominant trait. The trait that remains hidden is the recessive trait.

Parent (P):

$$TT \ \times \ tt$$

Pure Tall $\times$ Pure Dwarf

	T	T
t	Tt	Tt
t	Tt	Tt

Offspring (F$_1$):

$$Tt$$

All hybrid tall

Law of dominance—
all offspring are tall

Law of Segregation

The **law of segregation** states that during the formation of gametes, the two traits carried by each parent separate. Figure 6.1 shows this.

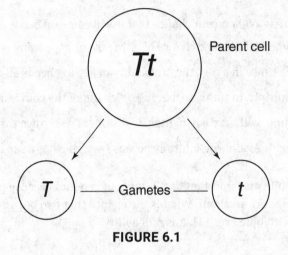

FIGURE 6.1

The cross that best exemplifies this law is the monohybrid cross (see the next section). In the monohybrid cross, a trait that was not evident in either parent can appear in the F_1 generation.

Monohybrid Cross

A **monohybrid cross** is a cross between two organisms that are each hybrid for a single trait, such as $Tt \times Tt$ (T = tall; t = dwarf). The **phenotype** (what the organism looks like) ratios that result from this cross are 3 tall to 1 short, or 75 percent tall plants to 25 percent dwarf plants. The **genotype** (type of genes) ratio is 25 percent homozygous dominant (TT) to 50 percent heterozygous (Tt) to 25 percent homozygous recessive (tt) or 1:2:1. These percentages are always true for a monohybrid cross, and you should memorize them.

HELPFUL HINT

If the phenotype ratio of the offspring is 3 to 1, both parents are hybrid for the trait.

F_1: $Tt \times Tt$

	T	t
T	TT	Tt
t	Tt	tt

Monohybrid cross

F_2: TT, Tt, or tt

Backcross or Testcross

The **testcross** or **backcross** is a way to determine whether an individual plant or animal showing the dominant trait is actually homozygous dominant (*BB*) or heterozygous (*Bb*).

To determine the genotype, the individual of unknown genotype (*B_*) is crossed with a homozygous recessive individual (*bb*). The genotype *B_* means that one allele is dominant (*B*) but the other allele is uncertain, so the individual is *BB* or *Bb*.

If the individual being tested is homozygous dominant (*BB*), all offspring of the testcross will show the dominant trait and have the hybrid (*Bb*) genotype. If the individual being tested is actually hybrid (*Bb*), we can expect that $\frac{1}{2}$ the offspring, or at least one individual, will show the recessive trait. Therefore, *if any offspring show the recessive trait, the parent of unknown genotype must be hybrid.*

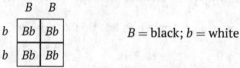

	B	B
b	Bb	Bb
b	Bb	Bb

B = black; *b* = white

If the parent of unknown genotype is *BB*, there can be no white offspring.

	B	b
b	Bb	bb
b	Bb	bb

B = black; *b* = white

If the parent of unknown genotype is hybrid, there is a 50 percent chance that any offspring will be white.

Law of Independent Assortment

The **law of independent assortment** applies when a cross is carried out between two individuals that are hybrid for two traits on separate chromosomes. This law states that during gamete formation, the genes for one trait (such as height, *T* or *t*) are not inherited along with the genes for another trait (such as seed color, *Y* or *y*). This example will use the following traits: *T* = tall; *t* = dwarf; *Y* = yellow seed; *y* = green seed.

Figure 6.2 represents a dihybrid individual (*TtYy*) where *T* will be inherited along with *Y*, while *t* will be inherited along with *y*. The trait for tall (*T*) will be separated from the trait for dwarf (*t*), and the trait for yellow seeds (*Y*) will be separated from the trait for green seeds (*y*).

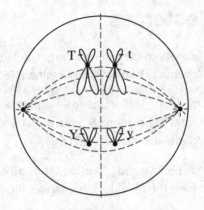

FIGURE 6.2

The only factor that determines how these alleles are inherited is how the homologous pairs (in this case *TY* and *ty*) happen to line up in metaphase of meiosis I, which is a random event.

During metaphase I, if the homologous pairs happen to line up like this:

$$
\begin{array}{c|c}
T & t \\
Y & y
\end{array}
$$

they will produce these gametes.   *TY*   *ty*

Instead, if the homologous pairs happen to line up like this:

$$
\begin{array}{c|c}
T & t \\
y & Y
\end{array}
$$

they will produce these gametes.   *Ty*   *tY*

Now follow a cross that demonstrates the law of independent assortment, a dihybrid cross.

Crossing *Tt Yy* × *Tt Yy* is called a dihybrid cross because it is a cross between individuals that are hybrid for two different traits—in this case, height and seed color. This cross can produce four different types of gametes: *TY, Ty, tY,* and *ty.* The following figure shows how to set up the Punnett square for this cross.

Gametes of one parent   ↓   *TY*   *Ty*   *tY*   *ty*   ← Gametes of other parent

	TY	*Ty*	*tY*	*ty*
TY	TTYY	TTYy	TtYY	TtYy
Ty	TTYy	TTyy	TtYy	Ttyy
tY	TtYY	TtYy	ttYY	ttYy
ty	TtYy	Ttyy	ttYy	ttyy

} Offspring

As you can see, there are many different genotypes possible in the offspring, but you should not pay attention to them. However, the phenotype ratios of the dihybrid cross are significant. The **phenotype ratio** from a dihybrid cross is **9:3:3:1**, 9 tall yellow, 3 tall green, 3 short yellow, and 1 short green. When shown as probabilities, this is:

F_2: 9/16 tall yellow   3/16 tall green   3/16 short yellow   1/16 short green

Beyond Mendelian Inheritance

Mendelian principles apply to traits determined by a single gene for which there are only two alleles. An example is height in pea plants (a single gene), which has only two alleles (tall and short). Now we will consider situations in which two or more genes are involved in determining a particular phenotype.

Incomplete Dominance

Incomplete dominance is characterized by blending. Here are two examples.

1. A long watermelon (*LL*) crossed with a round watermelon (*RR*) produces all oval watermelons (*RL*).
2. An animal with black fur (*BB*) crossed with one with white fur (*WW*) produces all gray fur (*BW*) animals.

Since neither trait is dominant, the convention for writing the genes uses different capital letters.

A red Japanese four o'clock flower (*RR*) crossed with a white Japanese four o'clock flower (*WW*) produces all pink offspring (*RW*).

Punnett square   *RR* × *WW*

	R	*R*
W	*RW*	*RW*
W	*RW*	*RW*

If two pink four o'clocks are crossed, there is a 25 percent chance that the offspring will be red (*RR*), a 25 percent chance the offspring will be white (*WW*), and a 50 percent chance the offspring will be pink (*RW*).

Punnett square   *RW* × *RW*

	R	*W*
R	*RR*	*RW*
W	*RW*	*WW*

Codominance

In **codominance**, both traits show. A good example is the *MN* blood group in humans (this is not related to the *ABO* blood group). There are three different blood types: *M*, *N*, and *MN*. These types are based on two distinct molecules located on the surface of the red blood cells. A person can be homozygous for one type of molecule (*MM*), homozygous for the other molecule (*NN*), or hybrid and have both molecules (*MN*) on the surface of their red blood cells. The *MN* genotype is not intermediate between *M* and *N*; both *M* and *N* traits are fully expressed on the surface of red blood cells, as shown in Figure 6.3.

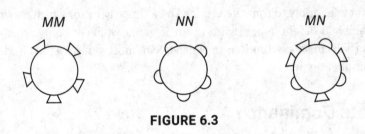

FIGURE 6.3

Multiple Alleles

Most genes in a population exist in only two allelic forms. For example, pea plants can be either tall (*T*) or dwarf (*t*). When there are more than two allelic forms of a gene, we refer to that situation as **multiple alleles**. In humans there are four different blood types (A, B, AB, and O) determined by the presence of specific molecules on the surface of the red blood cells. These four different blood types are determined by three alleles, *A*, *B*, and *O*. *A* and *B* are codominant and are often written as I^A and I^B. (*I* stands for immunoglobin.) When both alleles are present, they are both expressed, and the person has *AB* blood type. In addition, *O* is a recessive trait and is often written as *i*. A person can have any one of the six blood genotypes shown in Table 6.1.

TABLE 6.1 Human Blood Types and Genotypes

Blood Type	Genotype	
A	Homozygous *A*:	$I^A I^A$
A	Hybrid *A*:	$I^A i$
B	Homozygous *B*:	$I^B I^B$
B	Hybrid *B*:	$I^B i$
AB	Heterozygous:	$I^A I^B$
O	Homozygous *O*:	*ii*

Polygenic Inheritance

Many characteristics, such as skin color, hair color, and height, result from a blending of several separate genes that vary along a continuum. They are controlled by several genes and are called **polygenic**. Two parents who are short carry more genes for shortness than for tallness. However, they can have a child who inherits mostly genes for tallness from both parents and who will be taller than his/her parents. This wide variation in genotypes always results in a bell-shaped curve in an entire population, as seen in Figure 6.4, which shows another example of polygenic inheritance in the distribution of skin pigmentation across a population.

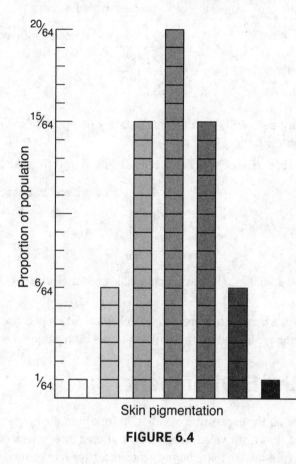

FIGURE 6.4

Sex-Linked Genes

Genes located on either sex chromosome are called **sex-linked genes**. Since there are only 55 genes on the Y chromosome (Y-linked) and since these are not passed on to females, we will consider only X-linked genes here (those genes on the X chromosome). There are around 900 X-linked genes, and they code for a wide range of characteristics in both males and females. Females (XX) inherit two copies of sex-linked genes. If a sex-linked trait is due to a recessive mutation, a female will express the phenotype only if she carries two mutated genes (X^aX^a). If she carries only one mutated X-linked gene, she will be a carrier (X^AX^a). Males (XY) inherit only one X-linked gene. As a result, if the male inherits a mutated X-linked gene (X^aY), he will express the gene. Although dominant sex-linked traits exist, recessive sex-linked traits are much more common. So, males suffer with sex-linked conditions more often than females do.

Here are some important facts about sex-linked traits.

- Common examples of recessive sex-linked traits are **color blindness** and **hemophilia**.
- All daughters of affected fathers are carriers. (See shaded squares.)

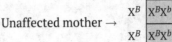

Unaffected mother →

- Sons cannot inherit a sex-linked trait from the father because the son inherits the Y chromosome from the father.
- A son has a 50 percent chance of inheriting a sex-linked trait from a carrier mother.

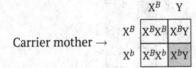

Carrier mother →

- There is no carrier state for X-linked traits in males. If a male has the gene, he will express it.
- It is uncommon for a female to express a sex-linked condition because in order to be affected, she must have inherited a mutant gene from both parents.

> **REMEMBER**
> In sex-linked inheritance, the father passes the trait to his daughters only.

Genes and the Environment

The environment can alter the expression of genes. In fruit flies, the expression of the mutation for vestigial wings (short, shriveled wings) can be altered by temperature. When raised in a hot environment, fruit flies that are homozygous recessive for vestigial wings can grow wings almost as long as normal wild-type wings. In humans, the development of intelligence is the result of an interaction of genetic predisposition and the influence of the environment.

Sex-Influenced Inheritance

Inheritance can be influenced by the sex of the individual carrying the traits. An example can be seen in male pattern baldness in humans, where hair is very thin on top of the head. This is not a sex-linked trait but, rather, a **sex-influenced trait**. Males and females express the gene for pattern baldness differently, as seen in Table 6.2.

TABLE 6.2 Sex Influence on Pattern Baldness

| Genotype | Phenotype | |
	Female	Male
BB	Bald	Bald
Bb	Not bald	Bald
bb	Not bald	Not bald

B = bald; *b* = not bald

Karyotype

A **karyotype** is a laboratory procedure that analyzes the size, shape, and number of chromosomes. Specialists prepare and photograph chromosomes during metaphase of mitosis when the chromosomes are fully condensed. Of the 46 human chromosomes, there are 44 (22 pairs) **autosomes** and 2 **sex chromosomes**. Figure 6.5 shows a karyotype of a normal male. Notice the X and Y sex chromosomes. Females, in contrast, have XX.

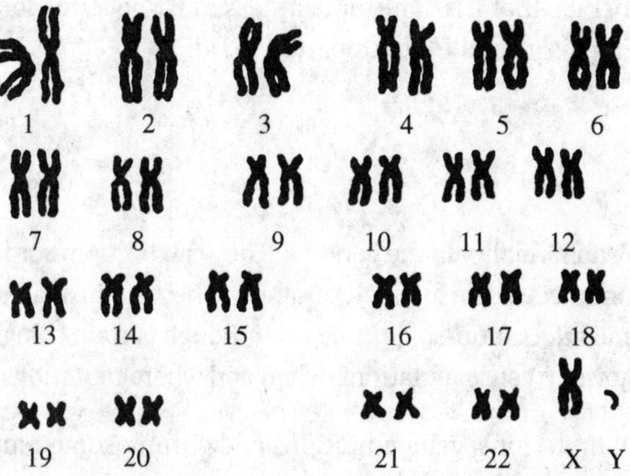

FIGURE 6.5 Karyotype of normal male

The Pedigree

A **pedigree** is a family tree that indicates the phenotype of one trait being studied for every member of a family. Geneticists use the pedigree to determine how a particular trait is inherited. By convention, females are represented by a circle and males by a square. The carrier state is not always shown. If it is, it is sometimes represented by a half-shaded shape. A shape is completely shaded in if a person exhibits the trait.

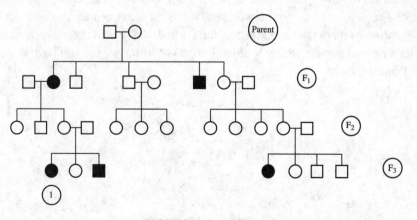

FIGURE 6.6 Pedigree

The pedigree in Figure 6.6 shows three generations of deafness. Try to determine the pattern of inheritance. (The shaded individuals are deaf.) First, eliminate any possibilities. The trait cannot be dominant (either sex-linked or autosomal), because in order for a child to have the condition, she or he would have had to have received one mutant gene from one afflicted parent, and nowhere is that the case. (All afflicted children have unaffected parents.) Also, the trait is not sex-linked recessive because in order for F_3 generation daughter #1 to have the condition, she would have had to inherit two mutant traits (X^dX^d), one from each parent. Yet her father does not have the condition. Since you have eliminated all the possibilities it could NOT be, you must conclude that the trait for deafness must be autosomal recessive. So, the genotype for daughter #1 is dd and for each parent is Dd.

Mutations

Mutations refer to any abnormality in the genome. They can occur in somatic (body) cells and be responsible for the spontaneous development of cancer. They can instead occur during gameto-genesis in germ cells and affect future offspring. Even though certain things like radiation and some chemicals are known to cause mutations, when and where mutations occur is random.

There are two types of mutations, gene mutations and chromosome mutations.

Gene mutations are caused by a change in the DNA sequence. Some human genetic disorders caused by gene and chromosome mutations are listed and described in Table 6.3. The nature of gene mutations at the DNA level is discussed in the next chapter.

Chromosome mutations can be observed under a light microscope. A chromosome may sustain a deletion or an addition, or a cell may have an entirely extra chromosome, which results from nondisjunction.

Nondisjunction

Nondisjunction is an error that sometimes happens during meiosis in which homologous chromosomes fail to separate as they should. This is illustrated in Figure 6.7. When this happens, one gamete receives two homologues, while the other gamete receives none. The remaining chromosomes may be unaffected and normal. If either of these aberrant gametes unites with a normal gamete during fertilization, the resulting zygote will have an abnormal number of chromosomes.

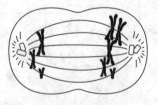

FIGURE 6.7 Nondisjunction

Any abnormal chromosome condition is known as **aneuploidy**. If a chromosome is present in triplicate, the condition is known as **trisomy**. People with Down syndrome have three copies of the 21st chromosome. The condition is referred to as trisomy-21.

An organism in which the cells have an extra set of chromosomes is referred to as triploid (3*n*). The cells of the endosperm or cotyledon of a seed are triploid. An organism with more than 3 sets of chromosomes is referred to as **polyploid**. Scientists breed plants to be polyploid because these plants will produce abnormally large flowers and fruit.

Human Inherited Disorders

Human genetic defects can be caused by either a gene or a chromosome mutation. (The cause of a gene mutation is discussed in the next chapter.) *Gene mutations are not visible under a microscope*, but chromosomal mutations can be assessed by doing a procedure called a karyotype. A karyotype is carried out on cells from a developing fetus to scan for chromosomal abnormalities such as trisomy-21, Down syndrome.

Chromosomal aberrations include:

1. **Deletion.** A fragment lacking a centromere is lost during cell division.
2. **Inversion.** A chromosomal fragment reattaches to its original chromosome but in the reverse orientation.
3. **Translocation.** A fragment of a chromosome becomes attached to a nonhomologous chromosome.
4. **Polyploidy.** A cell or an organism has extra sets of chromosomes.
5. **Nondisjunction.** Homologous chromosomes fail to separate during meiosis.

TABLE 6.3 Gene and Chromosome Mutations

Genetic Disorder	Pattern of Inheritance	Description
Phenylketonuria (PKU)	Autosomal recessive	Inability to break down the amino acid phenylalanine. Requires elimination of phenylalanine from diet; otherwise serious mental retardation will result.
Cystic fibrosis	Autosomal recessive	The most common lethal genetic disease in the U.S.; 1 out of 25 Caucasians is a carrier. Characterized by buildup of extracellular fluid in the lungs, digestive tract, etc.
Tay-Sachs disease	Autosomal recessive	Onset is early in life and is caused by lack of the enzyme needed to break down lipids necessary for normal brain function. It is common in Ashkenazi Jews and results in seizures, blindness, and early death.
Huntington's disease	Autosomal dominant	A degenerate disease of the nervous system resulting in certain and early death. Onset is usually in middle age.
Hemophilia	Sex-linked recessive	Caused by the absence of one or more proteins necessary for normal blood clotting.
Color blindness	Sex-linked recessive	Red-green color blindness is rarely more than an inconvenience.

TABLE 6.3 (Continued)

Chromosomal Disorder	Pattern of Inheritance	Description
Down syndrome	47 chromosomes with trisomy-21 = an extra chromosome 21	Characteristic facial features, mental impairment ranges from mild to severe, prone to developing Alzheimer's and leukemia.
Klinefelter's syndrome	XXY 47 chromosomes = a male with an extra X chromosome	Have male genitals, but the testes are abnormally small and these men are sterile.

Molecular Genetics

Today, everyone knows that deoxyribonucleic acid (DNA) is the molecule of heredity. However, that fact has not always been widely accepted. Until the early 1950s, many scientists believed that proteins were the molecules that make up genes and constitute inherited material. The work of many brilliant scientists has transformed our knowledge of the structure and function of the DNA molecule and led to the acceptance of DNA as the molecule responsible for heredity.

The Search for Inheritable Material

Griffith (1927) discovered the natural phenomenon known as **bacterial transformation,** which is the ability of bacteria to alter their genetic makeup by absorbing foreign DNA molecules from other bacterial cells and incorporating the foreign DNA into their own. He worked with different strains of the bacterium *Diplococcus pneumoniae*, which cause pneumonia.

Avery, MacLeod, and **McCarty** (1944) published their classic findings that the molecule that Griffith's bacteria was transferring was, in fact, DNA. They provided direct experimental evidence that DNA is the genetic material.

> ### INTERESTING FACT
> Rosalind Franklin was a major contributor to the study of large biological molecules. *Photo 51,* the now famous image of the double helix, was taken using techniques pioneered in her lab. She died in 1958.

Hershey and **Chase** (1952) proved that DNA, not proteins, is the molecule of inheritance when they tagged bacteriophages (viruses that attack bacteria) with the radioactive isotopes ^{32}P and ^{35}S. The ^{32}P labeled the DNA of the phage viruses, while the ^{35}S labeled the protein coat of the phage viruses. Hershey and Chase found that when bacteria were infected with phage viruses, ^{32}P from the virus entered the bacterium and produced thousands of progeny. However, no ^{35}S entered the bacterium. This is shown in Figure 7.1.

Rosalind Franklin (1950–1953), continuing the work begun by Maurice Wilkins, carried out the X-ray crystallography analysis of DNA that showed DNA to be a double helix. Her work was critical to Watson and Crick in developing their now-famous model of DNA.

Watson and **Crick** received the Nobel Prize in 1962 for correctly describing the structure of DNA as a double helix.

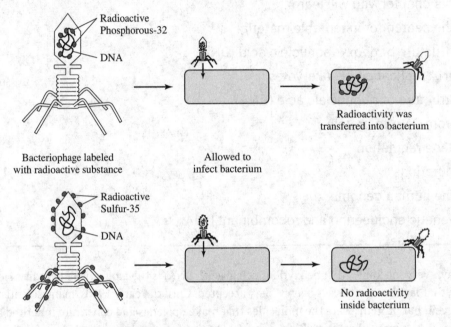

FIGURE 7.1 Hershey-Chase experiment

Meselson and **Stahl** (1953) proved Watson and Crick's hypothesis that DNA replicates in a semiconservative fashion. See Figure 7.2. They cultured bacteria in a medium containing heavy nitrogen (^{15}N) and then moved them to a medium containing light nitrogen (^{14}N), allowing the bacteria to replicate and divide once. The new bacterial DNA contained DNA consisting of one heavy strand and one light strand, thus proving Watson and Crick's theory.

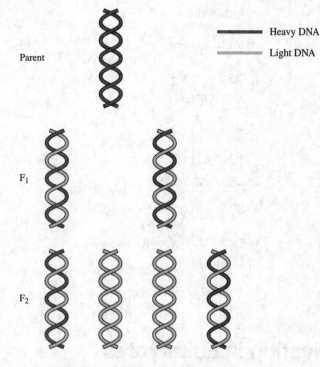

Parent

Heavy DNA
Light DNA

F_1

F_2

FIGURE 7.2 Semiconservative replication

Structure of Deoxyribonucleic Acid (DNA)

Figure 7.3 shows the DNA molecule.

- DNA is a **double helix** shaped like a twisted ladder.
- DNA consists of two complementary strands running in opposite directions from each other.
- It is a polymer made of repeating units called **nucleotides**.
- Each nucleotide consists of a 5-carbon sugar (deoxyribose), a phosphate molecule, and a nitrogenous base.
- Each nucleotide contains one of the four possible nitrogenous bases: **adenine** (**A**), **thymine** (**T**), **cytosine** (**C**), and **guanine** (**G**).
- A always bonds with T; C always bonds with G.
- The nucleotides of opposite chains are paired to one another by **hydrogen bonds**.

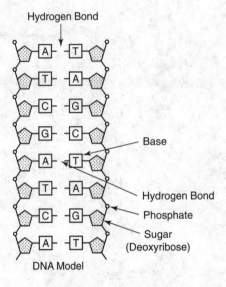

Hydrogen Bond

Base

Hydrogen Bond

Phosphate

Sugar
(Deoxyribose)

DNA Model

FIGURE 7.3 DNA molecule

DNA Replication in Eukaryotes

DNA replication is the making of an exact replica of DNA. The two new molecules of DNA that are produced each consist of one old strand and one new strand. This is called **semiconservative replication**, as proved by Meselson and Stahl. For a closer look at replication in eukaryotes, see Figure 7.4.

- Replication occurs during interphase in the life cycle of a cell.

- DNA polymerase catalyzes the replication of the new DNA.

- DNA polymerase also proofreads each new DNA strand, fixing errors and minimizing the occurrence of mutations.

- DNA unzips at the hydrogen bonds that connect the two strands of the double helix.

- Each strand of DNA serves as a template for the new strand according to the base-pairing rules: A with T and C with G.

- If a strand of DNA to be copied is   AAA TCG GAC,
  then the new strand is   TTT AGC CTG.

- Each time the DNA replicates, some nucleotides from the ends of the chromosomes are lost. To protect against the possible loss of genes at the ends of the chromosomes, some eukaryotic cells have special nonsense nucleotide sequences at the ends of chromosomes that repeat thousands of times. These protective ends of the chromosomes are called **telomeres**.

TIP
DNA replication occurs during interphase of the cell cycle.

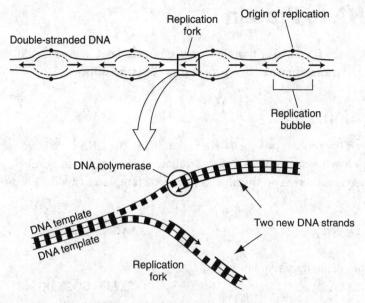

FIGURE 7.4 DNA replication in eukaryotes

Structure of Ribonucleic Acid (RNA)

- RNA is a single-stranded helix.

- It is a polymer made of repeating units called nucleotides.

- Each nucleotide consists of a 5-carbon sugar (ribose), a phosphate, and a nitrogenous base.

- Each nucleotide contains one of the four possible nitrogenous bases: **adenine** (**A**), **uracil** (**U**), **cytosine** (**C**), and **guanine** (**G**).

- There is no thymine in RNA. Uracil replaces thymine.

- There are three types of RNA: mRNA (messenger RNA), tRNA (transfer RNA), and rRNA (ribosomal RNA).

1. **Messenger RNA—mRNA** is made from a DNA template during a process called **transcription**. Messenger RNA carries the DNA code to the ribosomes, where the message is translated. Messenger RNA molecules are broken down after they carry out transcription.

2. **Transfer RNA—tRNA** is a clover leaf–shaped molecule that carries specific amino acid molecules to mRNA at the ribosome to help form a polypeptide during the process of **translation**.

3. **Ribosomal RNA—rRNA** is a structural molecule that makes up part of a ribosome. Each **ribosome** consists of a large and small subunit, each made of proteins and one or more rRNAs.

From DNA to Protein

The process whereby DNA makes proteins has been worked out in great detail. There are three main steps: **transcription**, **RNA processing**, and **translation**.

Transcription

Transcription is the process by which DNA makes RNA. The DNA code is transcribed into a **codon sequence** in messenger RNA (mRNA), following the base-pairing rules: A with U and C with G. Remember, there is no thymine in RNA. Uracil replaces thymine. Figure 7.5 shows transcription.

If the DNA sequence is: AAA TAA CCG GAC

Then the complementary sequence of
codons in mRNA is: UUU AUU GGC CUG

Transcription begins when the enzyme **RNA polymerase** binds to one strand of DNA at the **promoter region**. Next, RNA polymerase pries apart the two DNA strands and adds nucleotides to the growing end of the new strand, a process called *elongation*. When the newly forming RNA strand reaches a STOP codon (see Table 7.1), it **terminates** elongation and peels away from its DNA template, allowing the DNA double helix to re-form. Notice that whereas replication involved both strands of DNA, transcription involves only one of the DNA strands.

RNA Processing

After transcription but before the newly formed strand of RNA is shipped out of the nucleus to the ribosome, this **initial transcript** is edited or spliced by a series of enzymes, including those called snRNPs (pronounced "snurps"). The enzymes remove pieces of RNA that do not code for any protein. These *noncoding regions* that are removed are called **introns** (*intervening sequences*). The remaining portions, **exons** (*expressed sequences* or coding regions), are pieced back together to form the final transcript. As a result of this processing, the mRNA that leaves the nucleus is usually a great deal shorter than the piece that was initially transcribed.

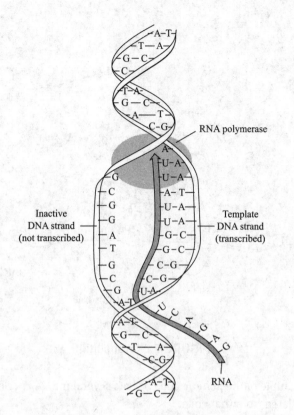

FIGURE 7.5 Transcription

Translation of mRNA into Protein

Translation is the process by which the mRNA sequence is converted into an amino acid sequence.

- Translation occurs at the ribosome.
- Amino acids present in the cytoplasm are carried by tRNA molecules to the codons of the mRNA strand at the ribosome according to the base-pairing rules (A with U and C with G).
- Several codons can code for the same amino acid. For example, codons UCU, UCC, UCA, and UCG all code for a single amino acid, serine.

Figure 7.6 shows the translation of the RNA code into an amino acid sequence.

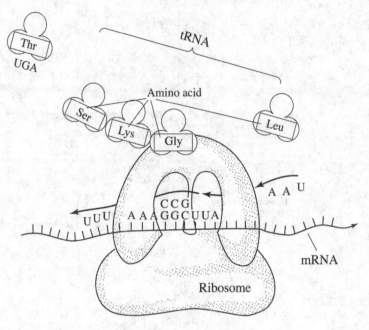

FIGURE 7.6 Translation

Table 7.1 shows a sample of the genetic code. It lists several mRNA codons and the amino acids that are translated from that code.

TABLE 7.1 Codons and Amino Acids

	SECOND BASE				
	U	**C**	**A**	**G**	
U	UUU ⎱ Phe UUC ⎰ UUA ⎱ Leu UUG ⎰	UCU ⎱ UCC ⎸ Ser UCA ⎸ UCG ⎰	UAU ⎱ Tyr UAC ⎰ UAA Stop UAG Stop	UGU ⎱ Cys UGC ⎰ UGA Stop UGG Trp	U C A G
C	CUU ⎱ CUC ⎸ Leu CUA ⎸ CUG ⎰	CCU ⎱ CCC ⎸ Pro CCA ⎸ CCG ⎰	CAU ⎱ His CAC ⎰ CAA ⎱ Gln CAG ⎰	CGU ⎱ CGC ⎸ Arg CGA ⎸ CGG ⎰	U C A G
A	AUU ⎱ AUC ⎸ Ile AUA ⎰ AUG Met or start	ACU ⎱ ACC ⎸ Thr ACA ⎸ ACG ⎰	AAU ⎱ Asn AAC ⎰ AAA ⎱ Lys AAG ⎰	AGU ⎱ Ser AGC ⎰ AGA ⎱ Arg AGG ⎰	U C A G
G	GUU ⎱ GUC ⎸ Val GUA ⎸ GUG ⎰	GCU ⎱ GCC ⎸ Ala GCA ⎸ GCG ⎰	GAU ⎱ Asp GAC ⎰ GAA ⎱ Glu GAG ⎰	GGU ⎱ GGC ⎸ Gly GGA ⎸ GGG ⎰	U C A G

FIRST BASE (5′ end)

THIRD BASE (3′ end)

By using the code in Table 7.1, determine what amino acid sequence will form if the DNA sequence is:

TAC AAA CAA GAA TCA

First, transcribe the DNA triplets into mRNA codons:

AUG UUU GUU CUU AGU

Then use the genetic code to translate the codon sequence to an amino acid sequence:

Methionine-Phenylalanine-Valine-Leucine-Serine

Illustrated Overview of Transcription, RNA Processing, and Translation

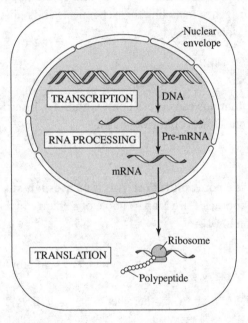

FIGURE 7.7 Overview of transcription, RNA processing, and translation

Gene Regulation

Every cell does not constantly synthesize every polypeptide it has the ability to make. For example, cells in the pancreas are not always producing tons of insulin because it is not always needed. That means that every gene in a cell is not turned on all the time. How does a cell know when to turn on a gene or when to turn it off? This is actually a very complex process in humans and one that is not understood well. However, a simple model for gene regulation can be found in bacteria in a region of DNA called the operon.

The **operon** is actually a cluster of functional genes plus the switches that turn them on and off. There are two types of operons. One is the *lac* or inducible operon, which is normally turned off unless it is actively induced or triggered to turn on by something in the environment.

The other is the repressible operon, which is always turned on unless it is actively turned off because it is temporarily not needed.

You may need to know some of the parts of the operon. The two most relevant ones are the promoter and the operator. The **promoter** is the binding site of RNA polymerase. RNA polymerase must always bind to DNA before transcription can take place. So, the promoter is like an "on" switch. The other important region is the **operator**. This is the binding site for the **repressor**, which turns off the *lac* operon. Another part of the operon, with a very funny name, is the **TATA box** (named for its sequences of alternating adenine and thymine), which helps RNA polymerase bind to the promoter.

Mutations

Mutations are changes in genetic material. They occur spontaneously and at random and can be caused by mutagenic agents, including toxic chemicals and radiation. Mutations are the raw material for natural selection.

Gene Mutations

Several types of gene mutations can occur: point mutations, insertions, and deletions. All types can have deleterious effects on the organism.

Point Mutation

The simplest mutation is a **point mutation**. This is a **base-pair substitution**, where one nucleotide converts to another. Here is an example of a change in an English sentence analogous to a point mutation in DNA:

Original		Point Mutation
↓		↓
THE FAT CAT SAW THE **D**OG.	*becomes*	THE FAT CAT SAW THE **H**OG.

The inherited genetic disorder sickle cell anemia results from a point mutation, like the one shown above, in the gene that codes for hemoglobin. The abnormal hemoglobin causes red blood cells to sickle when available oxygen is low. When red blood cells sickle, a variety of tissues may be deprived of oxygen and suffer severe and permanent damage.

It is possible, however, that a point mutation could result in a beneficial change for an organism or in no change in the proteins produced. Table 7.2 shows one example where a point mutation in DNA would result in no change in the amino acid sequence.

TABLE 7.2 Nonharmful Point Mutation

DNA	mRNA	Amino Acid Produced
AAA	UUU	Phenylalanine
AAG	UUC	Phenylalanine
Mutation ↑	Mutation ↑	No change in the amino acid

Insertion or Deletion

A second type of gene mutation results from a single nucleotide **insertion** or **deletion**. To continue the three-letter word analogy, a deletion is the loss of one letter and an insertion is the addition of a letter into the DNA sentence. Both mutations result in a **frameshift** because the entire reading frame is altered and is unreadable. Here are two examples.

Deletion of the Letter E

↓

THE FAT CAT SAW THE DOG → THF ATC ATS AWT HED OG

Insertion of the Letter T

↓

THE FAT CAT SAW THE DOG → THE FTA TCA TSA WTH EDO G

Depending on where in the DNA it occurs, a frameshift can have disastrous results. It can cause the formation of an altered polypeptide or no polypeptide at all.

Chromosome Mutations

Chromosome mutations are alterations in chromosome number or structure and are visible under a microscope. **Aneuploidy** is the term applied to having any abnormal number of chromosomes. One common example is Down syndrome, known officially as **trisomy-21**, in which a person is born with an extra chromosome 21. A karyotype of a male with an extra chromosome 21 (in box) is shown in Figure 7.8.

Having entire extra sets of chromosomes, such as 3*n* or 4*n*, is known as **polyploidy** and is common in plants. Polyploidy is responsible for unusually large and brilliantly colored flowers.

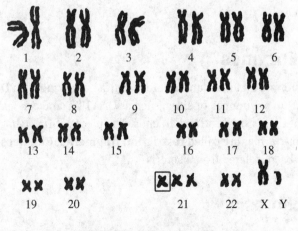

FIGURE 7.8 Trisomy-21

Aneuploidy and polyploidy both result from **nondisjunction**, where homologous pairs fail to separate during meiosis.

The Human Genome

The human **genome** (an organism's genetic material) consists of 3 billion base pairs of DNA and about 20,000 genes. Surprisingly, 97 percent of our DNA does NOT code for protein product. Some of this DNA consists of *regulatory sequences* that control gene expression. Some are *introns* that interrupt genes. Much of the DNA consists of *repetitive sequences* that may repeat 10 million times and never get transcribed. Some of the DNA are *pseudogenes*, former genes that have accumulated mutations over a long time. Scientists are only beginning to understand the makeup of DNA.

Genetic Engineering and Recombinant DNA

Recombinant DNA means taking DNA from two sources and combining them in one cell. The branch of science that uses recombinant DNA techniques for practical purposes is called **genetic engineering** or biotechnology. Two important areas of study in genetic engineering are gene therapy and environmental cleanup. Scientists are trying to learn how to insert functioning genes into cells to replace nonfunctioning ones in humans. If they are successful, it would mean an end to genetic diseases like cystic fibrosis and sickle cell anemia. Also, microbes are being engineered to degrade oil at oil spills or to decontaminate harmful chemicals at toxic mining sites or in water treatment plants.

On the downside, there is growing concern about the safety of genetically modified organisms (GMOs). Farmed salmon, for example, have been genetically modified by the addition of an extra growth hormone gene so they grow more quickly, and corn has been engineered to include foreign genes for resistance to herbicides and insect pests. One concern is that these genetically modified organisms will be released into the wild, thus spreading their engineered genes to wild species. Another is that these foreign genes might adversely affect people who eat the genetically engineered organisms. At present, governments are grappling with how to deal with these issues.

Restriction Enzymes

Restriction enzymes are an important tool for scientists working with DNA. They cut DNA at specific **recognition sequences** or **sites**, such as GAATTC, and can leave either blunt or "sticky" (a fragment that is single stranded) ends. The pieces of DNA that result from the cuts made by restriction enzymes are called **restriction fragments**. Hundreds of different restriction enzymes have been isolated from bacteria.

Gel Electrophoresis

Gel electrophoresis separates large molecules of DNA on the basis of their rate of movement as they flow through an agarose gel in an electric field. The smaller the molecule, the faster it runs through the gel. If necessary, the concentration of the agarose gel can be changed to provide a better separation of the tiny DNA fragments.

In order to run DNA through a gel, it must first be cut by restriction enzymes into pieces small enough to migrate through the gel. Once separated on a gel, the DNA can be analyzed in many ways.

This technique is the workhorse of any genetics lab.

Figure 7.9 shows an electrophoresis gel with four samples of DNA. The DNA in lanes 1, 2, and 4 were previously cut with restriction enzymes; the DNA in lane 3 was left uncut. Each sample is running in its own lane. The shorter pieces of DNA run farther and faster through the gel.

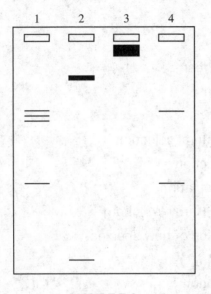

FIGURE 7.9

Lane 1 contains four bands of DNA, three larger pieces and one short piece. Lane 2 contains two pieces of DNA, one large and one tiny. Lane 3 contains one very large and uncut piece of DNA. Lane 4 contains two pieces of DNA.

Polymerase Chain Reaction

Devised in 1985, **polymerase chain reaction (PCR)** is a cell-free, automated technique by which a piece of DNA can be rapidly copied or amplified. Billions of copies of a fragment of DNA can be produced in a few hours. Once the DNA is amplified, these copies can be studied or used in a comparison with other DNA samples. This tool is a requirement for any genetics lab.

Evolution

Learning Objectives

In this chapter, you will learn:

- Evidence of evolution
- Lamarck vs. Darwin
- Darwin's theory of natural selection
- Types of natural selection
- Diversity within a population
- Population stability—Hardy-Weinberg
- Isolation and formation of new species
- Patterns of evolution
- Theories about evolution
- How life began
- The heterotroph hypothesis and the theory of endosymbiosis
- Mass extinctions
- Important concepts of evolution

REMEMBER
Individuals never evolve; only populations evolve.

An understanding of evolutionary theory helps scientists understand every field of biology from molecular biology to ecology. Evolution is the change in the genes of a population on Earth over time. **Microevolution** refers to the changes in one gene pool of a population over generations. **Macroevolution** refers to speciation, the formation of an entirely new species.

A critical thing to remember is that *individuals never change or evolve. A population is the smallest group that can evolve.* A **population** consists of all the members of one species in one place. For example, all the lions on the Masai plain in Kenya, or the blue jays in Madison, Wisconsin, are populations.

Evidence of Evolution

Six areas of scientific study that provide evidence for evolution are:

1. Fossil record
2. Comparative anatomy
3. Comparative biochemistry
4. Comparative embryology
5. Molecular biology
6. Biogeography

Fossil Record

The fossil record reveals the existence of species that have become extinct or have evolved into other species. The fossil record shows these important facts.

- Ninety-nine percent of all organisms that ever lived on Earth are now extinct.
- Through studies of radioactive dating and half-life, we know that Earth is about 4.6 billion years old.
- Prokaryotic cells are the oldest fossils and were the first organisms to develop on Earth.
- Paleontologists have discovered many **transitional fossils** that link older extinct fossils to modern species. For example, ***Archaeopteryx*** is a fossil that shows both reptile and bird characteristics. There also exist transition fossils that demonstrate that ***Hyracotherium*** (***Eohippus***), the ancient horse, is an ancestor of the modern horse, *Equus*.

Comparative Anatomy

Organisms that have similar anatomical structures are related to each other and share a common ancestor. For example, a comparison of dental structure in chimpanzees and humans demonstrates that we are related and that we both descended from a common ancestor less than 10 million years ago.

1. **Homologous structures.** Examples of homologous structures are the wing of the bat, the lateral fin of the whale, and the human arm, as shown in Figure 8.1. Although the function of each varies, they all have the same internal bone structure. The presence of homologous structure is evidence of common ancestry. The bat, whale, and human are all mammals. We *diverged* from a common ancestor millions of years ago. Homologous structures are examples of divergent evolution.

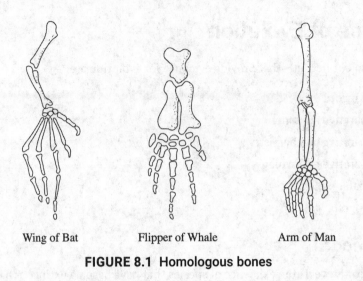

Wing of Bat   Flipper of Whale   Arm of Man

FIGURE 8.1 Homologous bones

2. **Analogous structures.** Analogous structures, such as a bat's wing and a fly's wing, have the same function but not the same underlying structure. The similarity is merely superficial and reflects adaptation to a similar environment. *These structures are not evidence of a common origin or common ancestry* but are evidence of convergent evolution.

3. **Vestigial structures.** Vestigial structures, such as the appendix, are evidence that the anatomy of animals has evolved. The appendix is a vestige of a structure needed when our ancient ancestors ate a very different diet.

Comparative Biochemistry

Organisms that have a common ancestor will have common biochemical pathways. The more closely related organisms are to each other, the more similar their biochemistry is. Humans and mice are both mammals. This close relationship is the reason medical researchers can test new medicines on mice and extrapolate the results to humans.

Comparative Embryology

Closely related organisms go through similar stages in their embryonic development because they evolved from a common ancestor. For example, all vertebrate embryos go through a stage in which they have gill pouches on the sides of their throats. In fish, the gill pouches develop into gills. In humans, they develop into eustachian tubes that connect the middle ear with the throat.

Molecular Biology

Since all aerobic organisms contain cells that carry out respiration and require electron transport chains, they also all contain the necessary polypeptide, cytochrome *c*. A comparison of the amino acid sequence of cytochrome *c* among different organisms shows which are most closely related. The cytochrome *c* in human cells is identical to that of our closest relative, the chimpanzee, but differs somewhat from that of a pig and is vastly different from the cytochrome *c* found in paramecia or in oak leaves.

Biogeography

The theory of **continental drift** states that about 250 million years ago, the continents were locked together in a single supercontinent known as **Pangaea**, which slowly separated into seven continents over the course of the next 150 million years. Study of the location of the fossils confirms the theory that marsupials migrated by land from South America across Antarctica to Australia before those two became separate continents about 55 million years ago. As a result most of the world's marsupials are isolated in Australia.

Lamarck vs. Darwin

Lamarck was a contemporary of Darwin and also developed a theory of evolution. His theory relied on the ideas of *inheritance of acquired characteristics* and *use and disuse*. Lamarck stated that individual organisms change in response to their environment. The giraffe developed a long neck because it ate leaves of the tall acacia tree for nourishment and had to stretch to reach them. Lamarck believed that the animals stretched their necks and passed on the acquired trait of an elongated neck to their offspring. Although this theory may seem funny today, it was widely accepted in the early 19th century.

Darwin was a naturalist and author who, when he was 22 in 1831, left England aboard the HMS *Beagle* to visit the Galapagos Islands, South America, Africa, and Australia. He collected, studied, and classified many organisms that no one in Europe had ever seen. His study of this amazing variety of organisms led him to develop his theory of natural selection, which explains how populations evolve and how new species develop. Darwin published *On the Origin of the Species* in 1859. It was a sensation. The first printing sold out on the first day it was published.

Darwin's Theory of Natural Selection

Here is the essence of Darwin's theory of natural selection.

- Populations tend to grow exponentially, to overpopulate, and to exceed their resources. Darwin developed this idea after reading **Malthus**, who published a treatise on population growth, disease, and famine in 1798.
- Overpopulation results in competition and a struggle for existence.
- In any population, there exists variation and an unequal ability of individuals to survive and reproduce. Although everyone would agree that this is obviously so, at the time, no one understood genetics and therefore could not explain the source of the variation. Mendel's theory of genetics, which was published in 1865 (although not understood for 35 years), would have given Darwin a basis for understanding variation in a population. Another source of variation, and a major one, mutation, was not understood until Hugo de Vries described it in the early 1900s.
- Only the best-fit individuals survive and pass on their traits to offspring. This is commonly known as **survival of the fittest**. According to Darwin, the degree of fitness is measured by the ability of an individual to survive and to reproduce in its environment. Evolution occurs as advantageous traits accumulate in a population.

> **TIP**
> The weakest part of Darwin's theory was his inability to explain the sources of diversity in a population.

How the Giraffe Got Its Long Neck

According to Darwin's theory, ancestral giraffes were short-necked animals, although neck length varied from individual to individual. As the population of animals competing for the limited food supply increased, the taller individuals had a better chance of surviving than those with shorter necks. Over time, the proportion of giraffes in the population with longer necks, which survived and reproduced, increased until only long-necked giraffes existed. Remember that no individual animal's neck grew longer. The average length of the neck in the *population* changed over time.

How the Peppered Moth Changed from Light to Dark

Until 1845 in England, most peppered moths were light colored; few dark individuals could be found. With increasing industrialization, smoke and soot polluted the environment, making all the plants and rocks black. By the 1950s, all moths in the industrialized regions were dark; few light-colored individuals could be found. Before the Industrial Revolution, white moths were camouflaged in their environment while dark moths were easy prey for predators (birds). After the environment was darkened by heavy pollution, things shifted. Dark moths were camouflaged and had the *selective advantage*. Within a relatively short time, 100 years, dark moths replaced the light moths in the population. This darkening due to industrialization is referred to as **industrial melanism**. Remember, no single individual moth changed. Instead, the frequency of an allele (for color) in the population changed.

Evolution and Drug Resistance

Not all evolution occurs slowly. Natural selection can produce very rapid shifts in populations. For example, only a few years after the discovery of antibiotics, bacteria appeared that were resistant to these drugs. **The appearance of antibiotics did not induce mutations for resistance; it merely killed susceptible bacteria**. Since only resistant individuals survived to reproduce, the next generation was resistant to the antibiotic they were exposed to. An entire population of bacterium can "become" resistant to a particular antibiotic in a matter of months. Once again, individual bacteria do not evolve; it is the population that evolves.

INTERESTING FACT

A new flu vaccine must be developed every year because the influenza virus mutates so rapidly.

The current treatment for AIDS (acquired immune deficiency syndrome) is a cocktail of drugs including AZT, which slows the progression of the disease. In some patients who have been taking the drugs for years, the virus that causes AIDS suddenly becomes resistant to these drugs and the patient quickly sickens. This is because the AIDS virus has the ability to mutate and evolve rapidly. The viruses that are susceptible to the drugs become inactivated, while those that have mutated are resistant. They survive and reproduce an entire population that is drug resistant. This is one reason we have not been able to cure AIDS.

Types of Natural Selection

The process of natural selection can alter the frequency of inherited traits in a population in three different ways, depending on which phenotypes in a population are favored. These types of selection are **stabilizing selection**, **diversifying** or **disruptive selection**, and **directional selection**.

Stabilizing Selection

Stabilizing selection eliminates the numbers of extremes and favors the more common intermediate forms. Many extreme forms are weeded out in this way. For example, stabilizing selection keeps the majority of birth weights in humans between 6 and 9 pounds. For babies much smaller or larger than this, mortality is greater. This is illustrated in Figure 8.2.

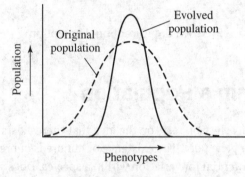

FIGURE 8.2 Stabilizing selection

Disruptive or Diversifying Selection

Disruptive selection increases the numbers of extreme types in a population at the expense of intermediate forms. A single population of snails can contain animals with either striped or plain shells. In the short term, this results in what is called *balanced polymorphism*, where two or more phenotypes coexist in a population. Over great lengths of time, disruptive selection may result in the formation of two entirely new species, as shown in Figure 8.3.

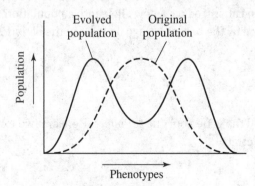

FIGURE 8.3 Disruptive selection

Directional Selection

Changing environmental conditions give rise to directional selection, where one phenotype replaces another in the gene pool. This was the case with peppered moths in England in the 20th century. Figure 8.4 diagrams this.

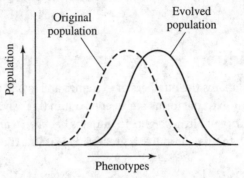

FIGURE 8.4 Directional selection

Diversity Within a Population

Darwin's theory of natural selection rests on the idea that there exists tremendous diversity or variation hidden in any gene pool. The existence of hundreds of breeds of dogs demonstrates the tremendous potential for variation within a species. Dogs such as the Great Dane, Chihuahua, and Beagle, which are all so very different from each other, belong to one species, *Canis familiaris*.

Although Darwin could not explain the origin of all the diversity he saw in different populations, we now understand the process. The sources of variation in a population are **mutation**, **genetic drift**, and **gene flow**.

Mutation

Mutations are changes in genetic material and are the raw material for evolutionary change. A single **point mutation** can introduce a new allele into a population. These mutations were first identified and named by the botanist Hugo de Vries in the early 1900s, who was studying polyploidy in plants.

Genetic Drift

Genetic drift is change in the gene pool due to *chance*. Here are two examples: the **bottleneck effect** and the **founder effect**.

The Bottleneck Effect

Natural disasters such as fire, earthquake, and flood reduce the size of a population nonselectively, resulting in a loss of genetic variation. The resulting population is much smaller and not representative of the original one. Certain alleles may be under- or overrepresented compared with the original population. This phenomenon is known as the bottleneck effect.

The Founder Effect

When a small population breaks away from a larger one to colonize a new area, it is most likely not genetically representative of the original larger population. Rare alleles may be overrepresented. This is known as the founder effect. It occurred in the Old Order of Amish of Lancaster, Pennsylvania, all of whom descended from a small group of settlers who came to the United States from Germany in the 1770s. Apparently, one or more of the settlers carried the rare but dominant gene for polydactyly, having extra fingers and toes. Due to the extreme isolation and intermarriage of the close community, this population now has a high incidence of polydactyly.

Gene Flow

Gene flow is the movement of alleles into or out of a population. It can occur as a result of the migration of fertile individuals or gametes between populations. For example, pollen from one valley can be carried by the wind across a mountain to another valley.

Population Stability— Hardy-Weinberg Equilibrium

Hardy and Weinberg, two scientists, developed a theorem that described a stable, nonevolving population—that is, one in which allelic frequency does not change. For example, if the frequency of an allele for a particular trait is 0.5 and the population is not evolving, then in 1,000 years, the frequency of that allele will still be 0.5. This is called Hardy-Weinberg equilibrium. According to Hardy-Weinberg, if the population is stable, the following must be true.

1. **The population must be very large.** In a small population, the smallest change in the gene pool will have a major effect in allelic frequencies. In a large population, a small change in the gene pool will be diluted by the sheer number of individuals and there will be no change in the frequency of alleles.

2. **The population must be isolated from other populations.** There must be no migration of organisms into or out of the gene pool because that could alter allelic frequencies.

3. **There must be no mutations in the population.** A mutation in the gene pool could cause a change in allelic frequency or introduce a new allele.

4. **Mating must be random.** If individuals select mates, then those individuals who are better fit will have a reproductive advantage and the population will evolve.

5. **There must be no natural selection.** Natural selection causes changes in relative frequencies of alleles in a gene pool.

Hardy-Weinberg Equation

The Hardy-Weinberg equation enables us to calculate frequencies of alleles in a population. Although it can be applied to complex situations of inheritance, for the purpose of explanation here, we will discuss a simple case—a gene locus with only two alleles. Scientists use the letter p to stand for the dominant allele and the letter q to stand for the recessive allele.

The Hardy-Weinberg equation is

$$p + q = 1 \quad \text{or} \quad p^2 + 2pq + q^2 = 1$$

Look at the monohybrid cross as a basis for this equation.

	A	a
A	AA	Aa
a	Aa	aa

$$p^2 = AA; \ 2pq = 2(Aa); \ q^2 = aa$$

TIP

p = dominant allele

q = recessive allele

Sample Problem

Take a pencil and do the problem along with the text.

If 9 percent of the population has blue eyes, what percent of the population is hybrid for brown eyes? Homozygous for brown eyes?

To solve this problem, follow these steps:

1. The trait for blue eyes is homozygous recessive, bb, and is represented by q^2. In this example, $q^2 = 9\%$.

   Convert to a decimal: $q^2 = 0.09$.

2. To solve for q, take the square root of $0.09 = 0.3$.
3. Since $p + q = 1$ and $q = 0.3$, then $p = 0.7$.
4. The hybrid condition is represented by $2pq$.
5. To solve for the percent of the population that is hybrid, substitute for $2(p)(q)$.

   $2(0.7)(0.3) = 0.42$.

   Convert to a percent: $2pq = 42\%$

6. Homozygous dominant is represented by p^2.

   $p^2 = (0.7)^2 = 0.49$

   Convert to a percent: $p^2 = 49\%$

TIP

p^2 represents the homozygous dominant individual.

q^2 represents the homozygous recessive individual.

$2pq$ represents all hybrid individuals.

BE CAREFUL

The square root of 0.09 is 0.3, not 0.03.

Isolation and New Species Formation

A **species** is a population whose members have the potential to interbreed in nature and produce viable, fertile offspring. For example, lions and tigers do not belong to the same species because, although they can be induced to interbreed in captivity, they would not naturally do so. Horses and donkeys do interbreed in nature. However, the offspring that results is a mule, which is not fertile. Therefore, the horse and donkey belong to different species.

Anything that fragments a population and isolates small groups of individuals may foster the formation of new species. This is true because isolated populations are subject to different selective pressures in their respective environments. If enough time elapses, the two populations may become so different that, even if they were brought back together, interbreeding would not occur. At this point, a new species is said to have come into being. Figure 8.5 shows two examples of isolating factors.

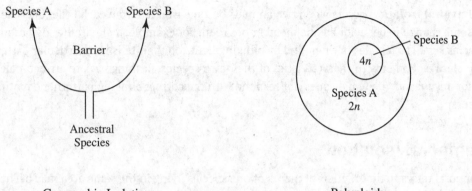

FIGURE 8.5

Six different forms of isolation commonly cause a new species to form:

1. Geographic isolation
2. Polyploidy
3. Habitat isolation
4. Behavioral isolation
5. Temporal isolation
6. Reproductive isolation

Geographic Isolation

Geographic isolation occurs when species are separated. Mountain ranges, canyons, rivers, lakes, or glaciers may cause significant isolation between species.

Polyploidy

Polyploidy is a type of mutation that results from errors during meiosis. Instead of being haploid (n) or diploid ($2n$), polyploid organisms can be tetraploid ($4n$) or octoploid ($8n$). Nearly half of all flowering plants and the vast majority of ferns are polyploid. Polyploid organisms cannot breed with organisms that are not polyploid and therefore are isolated from them. See Figure 8.5.

Habitat Isolation

Habitat isolation occurs when two organisms live in the same area but encounter each other rarely. Two species of one genus of snake can be found in the same geographic area, but one inhabits the water while the other is mainly terrestrial.

Behavioral Isolation

Behavioral isolation occurs when two animals become isolated from each other because of some change in behavior by one member or group. For example, male fireflies of various species signal to females of their kind by blinking lights on their tails in a particular pattern. Females respond only to characteristics of their own species, flashing back to attract males. If, for any reason, the female does not respond with the correct blinking pattern, no mating occurs.

Temporal Isolation

Temporal refers to time. Different plants of one species living in the same area may become functionally separated into two populations via temporal isolation because some plants become sexually mature earlier and begin to flower in the cooler part of the season while other plants flower in the later, warmer part of the growing season.

Reproductive Isolation

Closely related species may be unable to mate because of anatomical incompatibility. For example, a small male dog and a large female dog cannot mate because of the enormous size differences between the two animals.

Patterns of Evolution

How species evolve is classified into five patterns: divergent, convergent, parallel, coevolution, and adaptive radiation. These are illustrated in Figure 8.6.

Divergent Evolution

Divergent evolution occurs when a population becomes isolated (for any reason) from the rest of the species and becomes exposed to new selective pressures, causing it to evolve into a new species. Homologous structures are evidence of divergent evolution.

Convergent Evolution

When unrelated species occupy the same environment, they are subjected to similar selective pressures and show similar adaptations. The classic example of **convergent evolution** is the whale (a mammal) and the fish. Both have a streamlined appearance because that is advantageous in their environment. The underlying bone structure of the whale, however, reveals an ancestry common to mammals, not to fish. Analogous structures are evidence of convergent evolution.

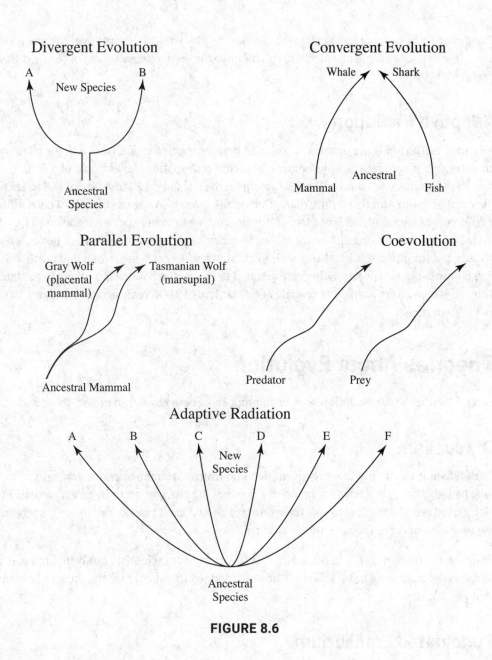

FIGURE 8.6

Parallel Evolution

Parallel evolution describes two related species that have made similar evolutionary adaptations after their divergence from a common ancestor. There are striking similarities between some placental mammals like the gray wolf of North America and the Tasmanian wolf, a marsupial, of Australia because they share a common ancestor and evolved in similar environments, thousands of miles apart.

Coevolution

Coevolution is the mutual evolutionary set of adaptations of two interacting species. Pollinator-plant relationships are one example. While feeding on the nectar from a flower, an insect, a bird, or a bat inadvertently ensures the reproductive success of the flower. A specific example is the honeybee that lives on the nectar of the Scottish broom flower.

The flower has a tripping mechanism that arches the stamens (male part) over the bee and dusts it with pollen, some of which will rub off onto the pistils (female part) of the next flower that the bee visits.

Adaptive Radiation

Adaptive radiation is the process by which numerous species evolve from a single common ancestor, introduced into an environment, reducing competition. An example of this phenomenon was made famous by Charles Darwin on the Galapagos Islands, about 600 miles off the coast of South America. While there, Darwin discovered 14 species of finches each filling a different ecological niche. Some live on the ground, while other species are adapted for life in trees. The most striking difference among the species is the variation in their beaks, which are adapted for different diets. Birds with thick, short beaks eat seeds, while birds with longer, pointy beaks eat insects. Others are adapted to eat cactus flowers and others to eat buds. They all evolved from a single ancestral species perhaps 10,000 years ago that radiated to fill 14 different niches.

Theories About Evolution

Several theories about evolution have been proposed. Some have been proven to be wrong.

Gradualism

Gradualism is the theory that organisms descend from a common ancestor gradually, over a long period of time, in a linear or branching fashion. Big changes occur by an accumulation of many small ones. According to this theory, fossils should exist as evidence of every stage in the evolution of every species with no missing links.

However, the fossil record is at odds with this theory because scientists rarely find transitional forms or missing links. Scientists have abandoned this theory for the theory of punctuated equilibrium.

Punctuated Equilibrium

The favored theory today is called **punctuated equilibrium**. It was developed by Stephen J. Gould and Niles Eldridge after they observed that Darwin's theory of gradualism was not supported by the fossil record. This theory proposes that new species appear relatively suddenly in the fossil record after long periods of little to no change. Most likely, a new species arises in a different place and expands its range, competing with and replacing the ancestral species that becomes extinct. Figure 8.7 illustrates both gradualism and punctuated equilibrium.

Spontaneous Generation

Spontaneous generation is the theory that living things emerge from nonliving or inanimate objects. This belief was disproved by two scientists. In the 17th century, Francesco Redi performed a now-famous experiment in which he put decaying meat into a group of wide-mouthed jars—some covered by lids, some covered by fine cheesecloth, and some left open. He demonstrated that maggots arose only where flies were able to lay their eggs.

In the early 1860s, Louis Pasteur, a microbiologist, used a goose-necked flask of his own making to prove that microorganisms appeared only as contaminants from the air and not spontaneously.

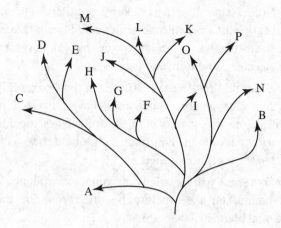

Gradualism

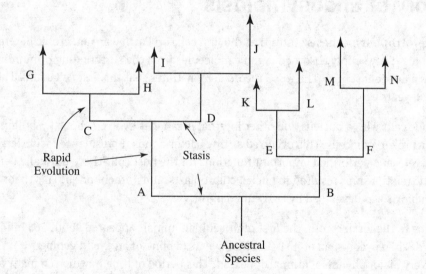

FIGURE 8.7 Punctuated equilibrium

How Life Began

All organisms alive today represent less than 1 percent of all the organisms that ever lived. This means that more than 99 percent of all life has gone extinct. The story of evolution based on scientific evidence goes like this.

Radioisotope dating of our solar system, including Earth, puts the age at about 4.6 billion years. After the big bang formed the planets, Earth's outer surface cooled and solidified to form a crust. The ancient environment most likely consisted of CH_4 (methane), NH_3 (ammonia), H_2O (vapor), and N_2 but lacked *free* oxygen. Intense heat, lightning, and UV radiation in the primitive atmosphere provided the energy for a multitude of chemical reactions that

REMEMBER
There was no free oxygen in the ancient Earth's atmosphere.

ultimately produced the first cell. Scientists have tried to mimic this early atmosphere to determine how the first organic molecules and earliest life developed. Here is a synopsis of those experiments.

1. A. I. Oparin and J. B. S. Haldane, in the 1920s, hypothesized separately that under the conditions of early Earth, organic molecules could form. They stated that in the absence of corrosively reactive molecular oxygen that would react with and degrade them, organic molecules could form and persist.

2. Stanley Miller and Harold Urey, in the 1950s, tested the Oparin-Haldane hypothesis and proved that almost any energy source would have converted inorganic molecules in the early atmosphere into a variety of organic molecules, including amino acids. They used electricity to mimic the lightning and UV light that must have been present in great amounts in the early atmosphere.

3. Sidney Fox, in more recent years, carried out similar experiments. He was able to produce membrane-bound, cell-like structures he called proteinoid microspheres, which would last for several hours in a laboratory.

The Heterotroph Hypothesis and the Theory of Endosymbiosis

REMEMBER
The first organisms on Earth were anaerobic heterotrophs.

The **heterotroph hypothesis** states that the first cells on Earth were anaerobic heterotrophic prokaryotes. They simply absorbed organic molecules from the surrounding primordial soup to use as a nutrient source. The fossil record demonstrates that the first cell evolved about 3.5 billion years ago.

Eukaryotic cells with a nucleus and other internal organelles evolved about 1.5 billion years ago from prokaryotic cells. This occurred as tiny bacteria took up residence inside larger prokaryotic cells and performed important functions for the host cell. These mutually beneficial symbiotic relationships resulted in nuclei, chloroplasts, and mitochondria. This theory of endosymbiosis was developed by Dr. Lynn Margulis.

According to the fossil record, the first multicellular animals appeared about 565 million years ago. Then over a span of just 40 million years (a blink of an eye in geologic terms), virtually every major phylum of animal appeared. This period of time in which so many animals appeared is known as the Cambrian explosion. As competition for limited resources increased in the oceans and as they evolved the traits necessary to live in a dry environment, animals and plants moved to land.

Several characteristics enabled animals to move to land:

- Lungs
- Skin to keep the animals from drying out
- Limbs to move about
- Mechanisms for internal fertilization
- Shell to protect their eggs and to keep them from drying out

Several characteristics enabled plants to move to land:

- Roots that anchor them into the soil and absorb water
- Supporting cells to enable them to compete favorably for light
- Vascular tissue to carry water upward
- Waxy molecule (cutin) to protect the leaves from dehydrating
- Seeds, a protective package for the embryo and its food

Mammals appeared about 210 million years ago. Primates, including apes, appeared about 25 million years ago. Humans did not evolve from apes; we both evolved from a common ancestor about 7 million years ago. Human ancestors arose in Africa. How they spread throughout the rest of the world is not agreed upon. However, scientists agree that modern humans, *Homo sapiens*, arose about 150,000 years ago.

Mass Extinctions

The fossil record shows that about 99% of every species that ever lived has gone extinct. A species may go extinct for many reasons, such as habitat destruction or drastic environmental change. There is currently a possible threat to the human species due to a worldwide increase in atmospheric CO_2 levels. In addition, even if physical factors (temperature, humidity, CO_2 levels) remain stable, in a world in which so many organisms are interdependent, the extinction of one species can result in the extinction of another species.

Although many small extinctions have occurred in Earth's history, there have been five major extinction events over the past 500 million years. These extinctions have been well documented in the fossil record. The two greatest mass extinctions are the **Permian extinction** (250 million years ago) and the **Cretaceous extinction** (65 million years ago).

The **Permian mass extinction** occurred during a period of enormous volcanic eruptions in what is now Siberia. Lava, hundreds of thousands of meters thick, covered an area the size of Europe. In addition, the eruptions emitted enough CO_2 into the atmosphere to cause the global climate to increase 6°C. Life on Earth was almost wiped out.

The **Cretaceous mass extinction** occurred when an asteroid 10 km wide crashed into the Yucatan Peninsula, MX. The scar can be seen from space. The current theory is that this collision caused a huge cloud of debris to billow into the atmosphere, blocking sunlight for months and the demise of much plant life. The loss of a food source resulted in the extinction of many marine and land animals including all the dinosaurs, excluding birds.

Important Concepts of Evolution

1. **Evolution is not always a slow process.** A population of bacteria can develop resistance to a particular antibiotic after only a few months of exposure in part because their generation time is so short.

2. **Evolution does not occur at the same rate in all organisms.** Human (*H. sapiens*) anatomy has changed a great deal since the divergence from *H. heidelbergensis* between 200,000 and 100,000 years ago. In contrast, the horseshoe crab has hardly changed at all when its anatomy is compared to fossils more than 400 million years old.

3. **Evolution does not always cause organisms to become more complex.** In at least two instances, populations of surface living tetra fish (*Astyanax mexicanus*) in Mexico have entered caves, become reproductively isolated, and lost their sight and pigmentation. In each case, the genetic changes affecting vision and pigmentation were different. Yet, the outcome was the same without the need to see in dark caves: fish without eyes survived and reproduced, creating populations of blind albino fish.

4. **Evolution occurs in populations, not individuals.** A single giraffe did not develop a long neck because it needed it. Instead, short-necked giraffes could not compete in a competitive environment and died out. Only giraffes with necks long enough to reach a food source survived and passed on their beneficial trait to the next generation.

5. **Evolution is directed by changes in the environment.** Whale fins are specially adapted for movement through the water. The humerus, radius, and ulna are present in both humans and whales. However, these bones are comparatively shorter in whales. The phalanges in whales are more numerous than in humans and are encased in soft tissue. In whales, the phalanges and soft tissue form a hydrofoil. However in humans, groups of two or three phalanges are separated from other phalanges to form fingers that are used for grasping.

Biological Diversity

Learning Objectives

In this chapter, you will learn:

o Three-domain classification system

o The four kingdoms of Eukarya: Protista, Fungi, Plantae, Animalia

o Evolutionary trends in animals

o Characteristics of animals

o Characteristics of mammals

o Characteristics of primates

o Cladograms or phylogenetic trees

Taxonomy is a system by which we name and classify all organisms, living and extinct. The system we use today is based on the system developed in the 18th century by Carl Linnaeus (Carl von Linné). It is known as the system of **binomial nomenclature** because every organism has a two-part name. For example, human is *Homo sapiens* and lion is *Panthera leo*. In addition, Linnaeus classified every organism into a hierarchy of **taxa**, or levels of organization. These taxa are **kingdom**, **phylum**, **class**, **order**, **family**, **genus**, and **species**. Kingdom is the most general, consisting of the most varied organisms, and species is the most specific, consisting of organisms that are the most similar.

In the 20th century, our system of classification went through many changes. In the 1950s and 1960s, all organisms were placed into only three kingdoms. From the 1960s to around 1990, scientists expanded the system to five kingdoms: Monera, Protista, Fungi, Plantae, and Animalia. In 1990, some scientists added a sixth kingdom, the Archaebacteria. This included **extremophiles**, microorganisms that live in extreme environments and that seemed so different from bacteria that they had to be placed into a separate kingdom.

Today, however, most scientists use another system, the **three-domain system**, which is based on DNA analysis. It more accurately reflects evolutionary history and the relationships among organisms. In our current system, all life is organized into three **domains**: Bacteria, Archaea, and Eukarya. These are superkingdoms that include four of the original kingdoms. (The kingdom **Monera** is no longer used because, in this system, prokaryotes are spread across two different domains, Archaea and Bacteria.)

The change to the three-domain system was necessary for one major reason: Archaea have so little in common with bacteria that they must have their own group. In addition, the name

> **REMEMBER**
> The term "Monera" is no longer used.

Archaebacteria had to be changed to Archaea because the Archaea are not bacteria, as seen in Figure 9.1. This sounds like a big change, but it is really simple.

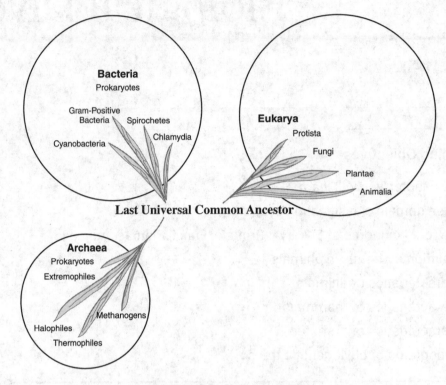

FIGURE 9.1 Three domains

The Three-Domain Classification System

All organisms are classified into one of the three domains. Notice that two domains, Bacteria and Archaea, both include prokaryotes. Notice also that the familiar kingdoms, Protista, Fungi, Plantae (plants), and Animalia (animals), are classified in the domain Eukarya.

Domain Bacteria

- Bacteria contain one double-stranded chromosome. They may also have extrachromosomal DNA in a plasmid.
- Many recombinant DNA techniques use plasmids to transform bacteria, allowing them to produce chemicals for human use.
- All are single-celled **prokaryotes** with no internal membranes (no nucleus, mitochondria, or chloroplasts).
- Some are anaerobes; some are aerobes.
- Bacteria play a vital role in the ecosystem as **decomposers** that recycle dead organic matter.
- Many are **pathogens**, disease causing.

- Bacteria play a vital role in **genetic engineering**. The bacteria from the human intestine, *Escherichia coli*, are used to manufacture human insulin.

- Some bacteria carry out **conjugation**, a primitive form of sexual reproduction where individuals exchange genetic material.

- Bacteria have a thick, rigid cell wall.

- Some are autotrophic (cyanobacteria); others are heterotrophic.

- Bacteria generally do not splice mRNA transcripts.

- Member species correspond roughly to the old grouping Eubacteria and include blue-green algae, bacteria like *E. coli* that live in the human intestine, those that cause disease like *Clostridium botulinum* and *Streptococcus*, and those necessary in the nitrogen cycle, like nitrogen-fixing bacteria, among others.

Domain Archaea

- Unicellular

- Prokaryotic—no internal membranes such as a nucleus

- Includes extremophiles, organisms that live in extreme environments, like

  1. **Methanogens.** Obtain energy in a unique way by producing methane from hydrogen

  2. **Halophiles.** Thrive in environments with high salt concentrations like Utah's Great Salt Lake

  3. **Thermophiles.** Thrive in very high temperatures, like in the hot springs in Yellowstone Park or in deep-sea hydrothermal vents

- Introns are present in some genes

Domain Eukarya

- All organisms have a nucleus and internal organelles.

- Eukarya includes the four remaining kingdoms: **Protista**, **Fungi**, **Plantae**, and **Animalia**.

> **INTERESTING FACT**
>
> All animals belong in the domain Eukarya.

The Four Kingdoms of Eukarya: Protista, Fungi, Plantae, Animalia

This section describes the key features of each of the four kingdoms. Although the kingdom Monera may still appear in some older textbooks, it is no longer in use. As previously described, the prokaryotes are now classified in two different domains, Bacteria and Archaea.

Kingdom Protista

- This kingdom includes the widest variety of organisms, but all are **eukaryotes**.

- Most are single celled, but many are primitive multicelled organisms.

- Includes **heterotrophs** and **autotrophs**.

- Examples of heterotrophs are **amoebas** and **paramecia**.

- Examples of autotrophs are **euglenas**, which have a red eyespot to locate light and chlorophyll to carry out photosynthesis.
- Protista move by various means: amoebas use pseudopods; paramecia use cilia; euglenas use a flagellum.
- Protista includes organisms that do not fit into the fungi or plant kingdoms, such as seaweeds and slime molds.
- Some protista (such as paramecia and algae) sometimes carry out **conjugation**, a primitive form of sexual reproduction where individuals exchange genetic material.
- Some cause serious diseases, like amoebic dysentery and malaria.

Kingdom Fungi

> **HELPFUL HINT**
>
> Fungi play an important role in ecology as decomposers.

- All are **heterotrophic eukaryotes**.
- They can be either unicellular or multicellular.
- Fungi carry out **extracellular digestion** by secreting hydrolytic enzymes outside the body. After digestion, the building blocks of the nutrients are absorbed into the body of the fungus by diffusion.
- They are important in the ecosystem as **decomposers**.
- Fungi are **saprobes**, organisms that obtain food from decaying organic matter. As such, they recycle nutrients in an ecosystem.
- Their cell walls are composed of **chitin**, not cellulose.
- Certain fungi combine with algae in a mutualistic, symbiotic relationship to form various lichens, which are photosynthetic. Lichens can survive harsh, cold environments and even live on bare rock. Lichens are often the **pioneer organisms**, the first to colonize a barren environment in an ecological succession.
- They reproduce asexually by budding (yeast), spore formation (bread mold), or fragmentation whereby a single parent breaks into parts that regenerate into whole new individuals.
- They also reproduce sexually.
- Examples include yeast, mold, mushrooms, and the fungus that causes athlete's foot.

Kingdom Plantae

- All are multicellular, nonmotile, autotrophic eukaryotes.
- Their cell walls are made of cellulose.
- Plants carry out photosynthesis using chlorophyll *a* and chlorophyll *b*.
- Plants store their carbohydrates as starch.
- They reproduce sexually by alternating between **gametophyte** (n) and **sporophyte** ($2n$) generations (a process known as alternation of generations).
- Some plants have vascular tissue (tracheophytes), and some have no vascular tissue (bryophytes).
- Examples include mosses, ferns, and both cone-bearing and flowering plants.

Kingdom Animalia

- All are heterotrophic, multicellular eukaryotes.
- Most are motile (can move on their own).
- Most animals reproduce sexually with a dominant diploid ($2n$) stage.
- In most species, a small flagellated sperm fertilizes a larger, nonmotile egg.
- The traditional way of classifying animals is primarily based on anatomical features (homologous structures) and embryonic development.
- They are grouped in 35 phyla, but we commonly discuss 9: porifera, cnidarians, platyhelminthes, nematodes, annelids, mollusks, arthropods, echinoderms, and chordates.

Evolutionary Trends in Animals

Organisms began as tiny, primitive, and single celled. They lived in the oceans. The first multicellular eukaryotes evolved about 1.5 billion years ago. The appearance of each phylum of animal represents the evolution of a new and successful body plan. These important trends include **specialization of tissues**, **germ layers**, **body symmetry**, **development of a head end**, and **body cavity formation**.

Specialized Cells, Tissues, and Organs

Begin with some definitions.

1. The **cell** is the basic unit of all forms of life. A neuron is a cell.
2. A **tissue** is a group of similar cells that perform one particular function. The sciatic nerve is a tissue.
3. An **organ** is a group of tissues that work together to perform related functions. The brain is a organ.

Sponges (porifera) consist of a loose federation of cells, which are not considered tissue because the cells are relatively unspecialized. They possess cells that can sense and react to the environment but have no real nerve or muscular tissue.

Cnidarians like the hydra and jellyfish possess only the most primitive and simplest forms of tissue.

As larger and more complex animals evolved, specialized cells joined to form real tissues, organs, and organ systems. **Flatworms** have organs but no organ systems.

More complex animals, like annelids (earthworms) and arthropods (grasshoppers), have organ systems.

Germ Layers

Germ layers are the main layers that form various tissues and organs of the body. They are formed early in embryonic development and include the ectoderm, endoderm, and mesoderm.

1. The **ectoderm**, or outermost layer, becomes the skin and nervous system, including the nerve cord and brain.
2. The **endoderm**, the innermost layer, becomes the viscera (guts) or the digestive system.
3. The **mesoderm**, or middle layer, becomes the blood, muscles, and bones.

Examples of animals with only two cell layers are porifera and cnidarians. Their bodies consist of ectoderm, endoderm, and **mesoglea** (middle glue), which holds the two layers together. The more complex animal phyla are **triploblastic**, having three true cell layers.

Body Symmetry

Whereas primitive animals exhibit **radial symmetry** (see Figure 9.3), sophisticated animals exhibit **bilateral symmetry** (see Figure 9.2). Echinoderms are an exception because they are an advanced phylum and exhibit bilateral symmetry as larvae but revert to radial symmetry as adults. In bilateral symmetry, the body is organized along a longitudinal axis with right and left sides that mirror each other. Most bilaterally symmetrical animals are triploblastic, with ectoderm, mesoderm, and endoderm.

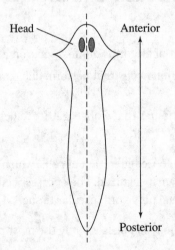

FIGURE 9.2 Bilateral symmetry in flatworm—planaria (advanced)

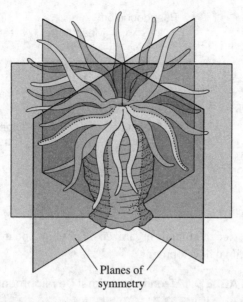

Planes of
symmetry

FIGURE 9.3 Radial symmetry in cnidarians (primitive)

Development of a Head (Cephalization)

Along with bilateral symmetry comes the development of a head end—the anterior, and of a rear end—the posterior. Sensory apparatus and a brain, or simply ganglia, are clustered at the anterior end. Digestive, excretory, and reproductive structures are located at the posterior end. This enables animals to move faster to flee or to capture prey successfully. Simple animals—sponges and cnidarians—do not have a head end. More sophisticated animals, beginning with flatworms and ending with chordates, all show cephalization.

Body Cavity Formation

The **coelom** is defined as a fluid-filled body cavity that is completely surrounded by mesoderm tissue as seen in Figure 9.4. It represents a significant advance in the course of animal evolution because it provides a space for elaborate organ systems, like the digestive tract or cardiovascular system. Primitive animals, like flatworms, do not have a coelom and are known as **acoelomates**. Their bodies are flat and all cells are in direct contact with their watery environment. Nematodes or roundworms are called **pseudocoelomates**; they have a fluid-filled tube between the endoderm and the mesoderm that acts as a **hydrostatic skeleton** to support the animal. **Coelomates** are animals with a coelom and are the most complex in the kingdom. These include the following phyla: Annelida, Mollusca, Arthropoda, and Chordata.

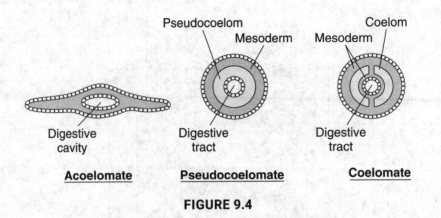

FIGURE 9.4

Figure 9.5 shows a family tree of animals organized by phylum and evolutionary trends. Table 9.1 identifies trends in animal development.

TABLE 9.1 Trends in Animal Development

From the Primitive	To the Complex
No symmetry or radial symmetry	Bilateral symmetry
No head	Head with sensory apparatus
Mesoglea holds two cell layers together	Three cell layers, including mesoderm
Acoelomate	Pseudocoelomate or coelomate
No true tissues	True tissues, organs, and organ systems
Little specialization	Much specialization
Sessile	Motile

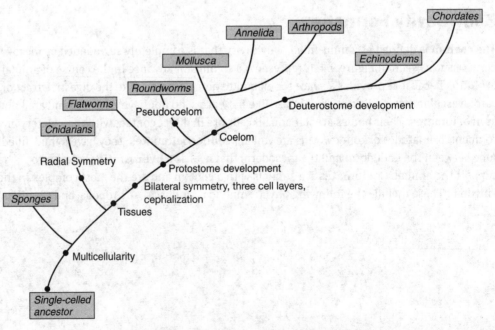

FIGURE 9.5 Trends in animal development

Characteristics of Animals

As you review the nine individual animal phyla, think in terms of strategies that animals have evolved to adapt to a particular environment. Also, notice the trends in development in animals from the primitive to the complex.

Porifera—Sponges

- No symmetry
- Have no nerve or muscle tissue, are **sessile**—do not move
- Filter nutrients from water drawn into a central cavity
- Consist of two cell layers only: ectoderm and endoderm connected by noncellular mesoglea
- Have specialized cells but no true tissues or organs; each cell carries out many functions
- Evolved from colonial organisms; if you squeeze a sponge through fine cheesecloth, it will separate into individual cells that will spontaneously reaggregate into a sponge
- Reproduce asexually by **fragmentation**
- Also reproduce sexually, are **hermaphrodites**

Cnidarians—Hydra and Jellyfish

- Radial symmetry
- Body plan is the **polyp** (vase shaped), which is mostly sessile, or **medusa** (upside-down bowl shaped), which is mostly **motile**
- Life cycle—some go through a planula larva (free-swimming) stage, then go through two reproductive stages: asexually reproducing (polyp) and sexually reproducing (medusa)
- Two cell layers only: ectoderm and endoderm connected by noncellular mesoglea
- Have a **gastrovascular cavity** where extracellular digestion occurs
- Carry out intracellular digestion inside body cells in **lysosomes**
- Have no transport system because every cell is in direct contact with the environment
- All members have stinging cells—**cnidocytes**—containing stingers, which are called **nematocysts**

Platyhelminthes—Flatworms Including Tapeworms

- They are the simplest animals with bilateral symmetry, an anterior end, and three distinct cell layers: ectoderm, endoderm, and mesoderm.
- The digestive cavity has only one opening for both ingestion and egestion, so food cannot be processed continuously.
- They have a solid body and have no room for true digestive or respiratory systems to circulate food or oxygen. Flatworms have solved this problem in a unique way. The body is so flat and thin that many body cells can exchange nutrients and wastes by diffusion with the environment.

Nematodes—Roundworms

- Nematodes are unsegmented worms with bilateral symmetry but little sensory apparatus.
- Many are parasitic. *Trichinella* causes trichinosis, which is contracted by eating uncooked pork.
- One species, *C. elegans,* is widely used as an animal model in studying genes and embryonic development.

Annelids—Segmented Worms Like Earthworms, Leeches

- Bilateral symmetry with little sensory apparatus
- Digestive tract is a tube-within-a-tube consisting of **crop**, **gizzard**, and intestine
- **Nephridia** for excretion of the nitrogen waste, urea
- **Closed circulatory system**—heart consists of five pairs of aortic arches
- Blood contains hemoglobin and carries oxygen
- Diffusion of oxygen and carbon dioxide through moist skin
- Hermaphrodites

Mollusks—Squids, Octopuses, Slugs, Clams, and Snails

- Have **soft body** often protected by a hard calcium-containing shell
- **Open circulatory system** with blood-filled spaces called **hemocoels** or **sinuses**
- Have bilateral symmetry with three distinct body zones:
  1. Head-foot, which contains both sensory and motor organs
  2. Visceral mass, which contains the organs of digestion, excretion, and reproduction
  3. Mantle, a specialized tissue that surrounds the visceral mass and secretes the shell
- Radula, a movable, tooth-bearing structure, acts like a tongue
- Most have gills and nephridia

Arthropods—Insecta (Grasshopper), Crustacea (Shrimp, Crab), Arachnida (Spider)

- Jointed appendages
- Segmented into head, thorax, and abdomen
- More sensory apparatus than in annelids, giving them more speed and freedom of movement
- Chitinous exoskeleton protects the animal and aids in movement
- **Open circulatory system** with a tubular heart and **hemocoels**, sinuses
- **Malpighian tubules** for removal of nitrogenous wastes, uric acid
- Air ducts called **trachea** bring air from the environment into hemocoels

Echinoderms—Sea Stars (Starfish) and Sea Urchins

- Most are sessile or slow moving.
- They have bilateral symmetry as an embryo but revert to the primitive radial symmetry as an adult. The radial anatomy of the adult is an adaptation to a sedentary lifestyle.
- Their water vascular system creates hydrostatic support for the tube feet, the locomotive structures.
- Echinoderms reproduce by sexual reproduction with external fertilization.
- They can also reproduce by fragmentation and regeneration. Any piece of a sea star that contains part of the central canal will form a completely new organism.
- Sea stars have an endoskeleton consisting of calcium plates. An endoskeleton grows with the body. In contrast, an exoskeleton does not and must be shed periodically.

Chordates—Fish, Amphibians, Reptiles, Birds, Mammals

- Chordates have a **notochord**, a rod that extends the length of the body and serves as a flexible axis.
- They have a dorsal, hollow nerve cord.
- The tail aids in movement and balance. The coccyx bone in humans is a vestige of a tail.
- Birds and mammals are **homeotherms**—they maintain a consistent body temperature. All other chordates—fish, amphibians, and reptiles—are cold-blooded, although some reptiles are **endotherms** (heat from within) and are able to raise their body temperature.

Characteristics of Mammals

> **HELPFUL HINT**
>
> Birds and mammals are both homeotherms. But birds are not mammals.

- Mammals belong to the phylum Chordata.
- Mothers nourish their babies with milk from mammary glands.
- They have hair or fur.
- Mammals are endotherms (warm-blooded).
- Most are placental mammals (eutherians)—the embryo develops internally in a uterus connected to the mother by a placenta, where nutrients diffuse from mother to embryo.
- Some, the marsupials, including kangaroos, are born very early in embryonic development. The "joey" completes its development while nursing in the mother's pouch attached to a teat.
- Monotremes, egg-laying mammals, like the duck-billed platypus and the spiny anteater, derive nutrients from a shelled egg.

Table 9.2 shows the classification of three mammals using our current three-domain system and taxonomy originally developed by Linnaeus.

TABLE 9.2 Sample Classification

Taxa	Human	Lion	Dog
Domain	Eukarya	Eukarya	Eukarya
Kingdom	Animalia	Animalia	Animalia
Phylum	Chordata	Chordata	Chordata
Class	Mammalia	Mammalia	Mammalia
Order	Primate	Carnivora	Carnivora
Family	Hominid	Felidae	Canidae
Genus	*Homo*	*Panthera*	*Canis*
Species	*sapiens*	*leo*	*familiaris*

Characteristics of Primates

Humans are primates. Primates descended from insectivores, probably from small, tree-dwelling mammals. Primates have dexterous hands and opposable thumbs, which make it possible to do fine-motor tasks. Nails have replaced claws. Hands and fingers contain many nerve endings and are sensitive. The eyes of a primate are forward facing and set close together. Front-facing eyes foster face-to-face communication. Close-set eyes are responsible for overlapping fields of vision, which enhance depth perception and hand-eye coordination. Although mammals devote much energy to the parenting of young, primates engage in the most intense parenting of any mammal. Primates usually have single births and nurture their young for a long time. Primates include humans, gorillas, chimpanzees, orangutans, gibbons, and the old world and new world monkeys.

Cladograms and Phylogenetic Trees

All living things evolved from the last universal common ancestor (LUCA) almost 4 billion years ago, which is why the principles of biology apply to all organisms. A **cladogram** or **phylogenetic tree** is a diagrammatic representation of that evolutionary history, and it is based on DNA sequences. To build a cladogram, you must distinguish the difference between **shared traits**, those that organisms have in common, and **derived traits**, new characteristics or innovations that are not shared with ancestors.

Here is a chart that shows the derived characteristics in several organisms.

(+ means the trait is present; − means the trait is absent)

Organism	Derived Characteristics		
	Backbone	Legs	Hair
Earthworm	−	−	−
Salmon	+	−	−
Lizard	+	+	−
Cat	+	+	+

Figure 9.6 is a cladogram based on the information presented in the chart.

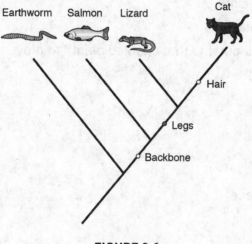

FIGURE 9.6

The derived traits (new characteristics)—backbone, legs, and hair—are placed along the base of the cladogram. You can see that the earthworm has none of these traits. The salmon, lizard, and cat all have a backbone. Only the lizard and cat have legs, and only the cat has hair. Deductions you can draw from this presentation are these:

> The cat and lizard are most closely related.
> The cat is less related to the salmon than to the lizard.
> The earthworm is the most primitive, and the cat is the most advanced.
> There is no information about when these derived traits evolved or when the animals' ancestors diverged from each other.

If you were asked to provide a shared trait for all the animals shown including the earthworm, what would you say? There are many choices, but a good one would be:

Answer: bilateral symmetry.

Plants

Learning Objectives

In this chapter, you will learn:

- Classification of plants
- A cladogram for plants
- Evolutionary developments that enabled plants to move to land
- How plants grow
- Roots
- Stems
- The leaf
- Stomates
- Types of plant tissue
- Transport in plants
- Plant reproduction
- Alternation of generations
- Plant responses to stimuli

Plants include all multicelled, eukaryotic, photosynthetic **autotrophs**. Their cell walls are made of **cellulose**, and they store carbohydrates as **starch**. Biologists believe that modern, multicelled plants evolved from the green algae Chlorophyta that lived in freshwater.

Classification of Plants

Plants can be classified as either bryophytes or tracheophytes.

Bryophytes

- Bryophytes are primitive plants that lack vascular tissue.
- They must live in moist environments because they have no roots or xylem and must absorb and transport water by osmosis.
- Bryophytes are tiny because they lack the lignin-fortified tissue necessary to support tall plants on land.
- Bryophytes include mosses, liverworts, and hornworts.

Tracheophytes

- Tracheophytes have transport vessels—xylem and phloem.
- They include ancient seedless plants, like ferns, that reproduce by spores.
- They include modern plants that reproduce by seeds.
- Those with seeds are further subdivided into **gymnosperms** and **angiosperms**.

Gymnosperms

Gymnosperms are conifers, the cone-bearing plants that produce seeds on the surface of cones. They have various modifications to make them more resistant to wind, cold, and drought. These include needle-shaped leaves, a thick and waxy cuticle, and stomates located in stomatal crypts to reduce water loss even further. Cedars, sequoias, redwoods, pines, yews, and junipers are all gymnosperms.

Angiosperms

Angiosperms are flowering plants in which seeds develop inside ovaries of flowers. After pollination, the ovary becomes the *fruit*. They are the most diverse and plentiful plants on Earth. Roses, daisies, fruits, nuts, grains, and grasses are just a few of the examples of angiosperms. These plants are subdivided into two more groups: **monocotyledons** (monocots) and **eudicots**, which include complex flowering plants called **dicotyledons** (dicots). Table 10.1 shows the main differences between monocots and dicots.

TABLE 10.1 The Principal Differences Between Monocots and Dicots

Characteristic	Monocots	Dicots
Cotyledons (seed leaves)	One	Two
Vascular bundles in stem	Scattered	In a ring
Leaf venation	Parallel	Netlike
Floral parts	Usually in 3s	Usually in 4s or 5s
Roots	See Figure 10.2	

Examples of monocots are the grasses: wheat, corn, oats, lawn grass, and rice. Monocots provide the food for most of the world. Palm trees are also monocots. Examples of dicots are daisies, roses, carrots, and most flowering plants you could name. Oak, walnut, cherry, and most other trees you could name are dicots.

> **TIP**
> In the dicot stem, vascular bundles are located in a ring around the edge.
>
>

> **TIP**
> In the monocot stem, vascular bundles are scattered across the stem.

A Cladogram for Plants

The cladogram in Figure 10.1 shows the evolutionary relationships among the four main groups of living plants using the presence or absence of three *derived traits*: vascular tissue, seeds, and flowers.

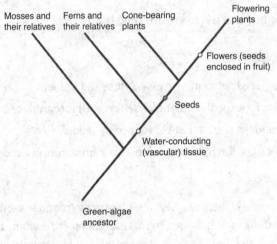

FIGURE 10.1

Evolutionary Developments That Enabled Plants to Move to Land

Plants began life in the seas and moved to land as competition for resources increased. The biggest problems a plant on land faces are supporting a plant body and both absorbing and conserving water. Several modifications enable plants to live on land.

- Cell walls made of cellulose support the plant whose cells, unsupported by a watery environment, must maintain their own shape.
- Roots and root hairs absorb water and nutrients from the soil.
- **Stomates** open to exchange photosynthetic gases and close to minimize excessive water loss.
- The waxy coating on the leaves, **cutin**, helps prevents excess water loss from the leaves.
- In some plants, gametes and zygotes form within a protective jacket of cells called **gametangia** that prevents drying out.
- **Sporopollenin**, a tough polymer, is resistant to almost all kinds of environmental damage and protects plants in a harsh terrestrial environment. It is found in the walls of spores and pollen.
- Seeds and pollen have a protective coat that prevents desiccation. They are also a means of dispersing offspring.
- Reduction of the primitive gametophyte (n) generation occurs.

How Plants Grow

Unlike animals, plants continue to grow as long as they live because plants have **meristem tissue** that continually divides, generating new cells. Plants grow in two ways: **primary growth** and **secondary growth**.

Primary Growth

Primary growth is vertical. It is the elongation of the plant down into the soil and up into the air. New cells arise from the constantly dividing growth layer called the **apical meristem**, which is located at the buds of shoots and the tips of the roots.

Root growth is concentrated near the root tip. Three zones of cells at different stages of primary growth are located there: the **zone of cell division** called the apical meristem, the **zone of elongation**, and the **zone of differentiation**. The root tip is protected by a root cap that secretes a substance that helps digest the earth as the root tip grows through the soil.

Figure 10.2 shows a longitudinal section of a root.

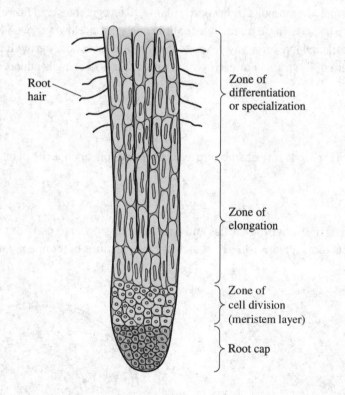

FIGURE 10.2 Root tip

Zone of Cell Division

This zone contains meristem cells that are actively dividing and are responsible for producing new cells that grow down into the soil. This is the region you observed under the microscope in the lab when you were studying cells undergoing mitosis.

Zone of Elongation

Cells in this zone elongate and are responsible for pushing the root cap downward and deeper into the soil.

Zone of Differentiation

Cells in this zone undergo specialization into three primary meristems that give rise to three tissue systems in the plant: the epidermis, ground tissue, and xylem and phloem.

Secondary Growth

Secondary growth means lateral growth or an increase in girth. New cells are provided by the lateral meristem. In herbaceous (nonwoody) plants, like vegetables and flowers, there is only primary growth because these plants live for only one season. However, woody plants are protected by bark and live for many years. In these plants, secondary growth is responsible for the enlargement of the trunk. For each year of growth, another ring is added.

Roots

The function of the roots is to absorb nutrients from the soil, anchor the plant, and store food.

Structure

The root consists of specialized tissues and structures organized to carry out the various functions of the roots. Figure 10.3 is a sketch of cross sections of monocot and dicot roots.

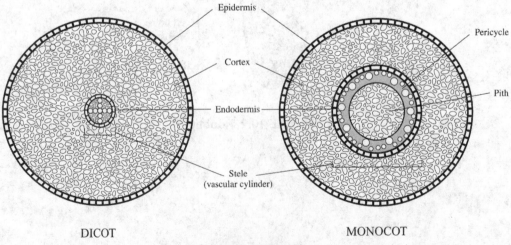

FIGURE 10.3 Roots

Epidermis

The epidermis covers the entire surface of the root and is modified for absorption. Slender cytoplasmic projections from the epidermal cells, called root hairs, extend out from each cell and greatly increase the root's absorptive surface area.

Cortex

The function of the cortex is storage. It consists of **parenchyma cells** that contain many **plastids** for the storage of starch and other organic substances.

Stele

The function of the vascular cylinder or stele is transport. It consists of vascular tissues (xylem and phloem) surrounded by one or more layers of tissue called the pericycle, from which lateral roots arise.

Endoderm

The vascular cylinder is surrounded by a tightly packed layer of cells called the endodermis. Each endoderm cell is wrapped with the **Casparian strip**, a continuous band of waxy material that is impervious to water and dissolved minerals. The function of the endoderm is to select what minerals enter the vascular cylinder and the body of the plant.

Absorption of Nutrients and Water

Plants use their roots to absorb nutrients and water from the soil. These then must be absorbed by the cells themselves.

Apoplast and Symplast

The movement of water and solutes across a plant, called lateral movement, is accomplished along the symplast and apoplast. The **symplast** is a continuous system of cytoplasm of cells interconnected by **plasmodesmata**. The **apoplast** is the network of cell walls and intercellular spaces within a plant body that permits extensive extracellular movement of water within a plant.

Mycorrhizae

In mature plants of many species where older regions of roots lack root hairs, **mycorrhizae** supply the plant with water and minerals. Mycorrhizae are symbiotic structures consisting of the plant's roots intermingled with the hyphae (filaments) of a fungus that greatly increase the quantity of nutrients that a plant can absorb.

Rhizobium

Rhizobium is a symbiotic bacterium that lives in the nodules on roots of specific legumes. It fixes nitrogen gas from the air into a form of nitrogen the plant requires.

Types of Roots

The **taproot** is a single, large root that gives rise to lateral branch roots. In many dicots, the primary root is the taproot. Some taproots "tap" water deep in the soil. Others, like carrots, beets, and turnips, are modified for storage.

A fibrous root system, common in monocots like grasses, holds the plant firmly in place. As a result, grasses make fine ground cover because they minimize soil erosion.

Adventitious roots are roots that arise aboveground. Here are two examples:

1. **Aerial roots.** Trees that grow in swamps or salt marshes like mangroves have aerial roots that stick up out of the water and serve to aerate the root cells. English ivy has aerial roots that enable the ivy to cling to the sides of buildings.
2. **Prop roots.** Some tall plants like corn have prop roots that grow aboveground out from the base of the stem and help support the plant.

Stems

The function of the stem is support. This support allows the leaves to receive the most light. Stems also transport water and minerals from the soil and nutrients from the leaves to the rest of the plant.

Structure

Vascular tissue runs the length of the stem in strands called vascular bundles. Each bundle contains xylem on the inside, phloem on the outside, and meristem tissue between the two. In monocots, the vascular bundles are scattered throughout the stem. In dicots, they are arranged in a ring around the edge of the stem. The ground tissue of the dicot stem consists of cortex and **pith**, parenchymal tissues modified for storage. In monocots, ground tissue is not partitioned into cortex and pith.

Figure 10.4 shows the difference between monocot and dicot stems.

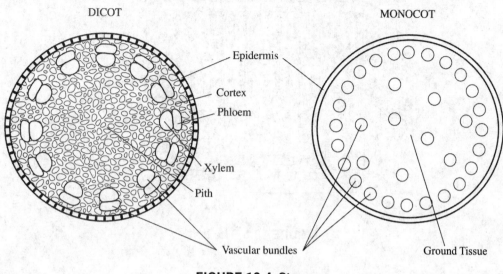

FIGURE 10.4 Stems

The Leaf

The leaf is organized to maximize sugar production while minimizing water loss. Table 10.2 lists the parts of a leaf and their functions, and Figure 10.5 illustrates them.

TABLE 10.2 Leaf Parts and Function

Part of Leaf	Function
Epidermis—upper and lower	Protection
Waxy cuticle—made of cutin	Minimizes water loss
Guard cells—modified epidermal cells, contain chloroplasts	Control the opening of the stomates
Palisade mesophyll—tightly packed	Photosynthesis
Spongy mesophyll—loosely packed	Photosynthesis Diffusion and exchange of gases into and out of these cells
Veins—located in the mesophyll	Carry water and nutrients from the soil to the leaves and carry sugar, the product of photosynthesis, from the leaves to the rest of the plant

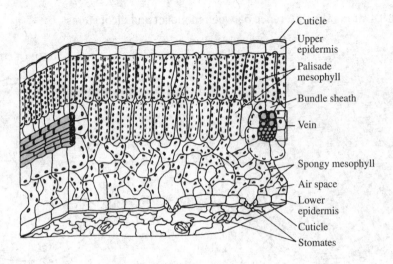

Cuticle
Upper epidermis
Palisade mesophyll
Bundle sheath
Vein
Spongy mesophyll
Air space
Lower epidermis
Cuticle
Stomates

FIGURE 10.5 The leaf

Stomates

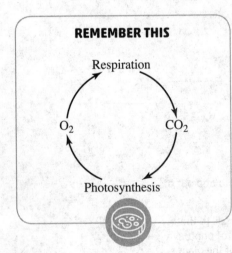

REMEMBER THIS

Respiration

O_2 CO_2

Photosynthesis

When plant cells carry out cellular respiration, they take in oxygen and give off carbon dioxide. When plant cells carry out photosynthesis, however, they take in carbon dioxide and give off oxygen and water vapor. Plants exchange these gases between air spaces in the spongy mesophyll and the exterior of the leaf by opening their stomates. So, why do plants ever close their stomates? If stomates were kept open all the time, the plant would lose so much water through **transpiration** (loss of water from the leaf) it could not survive. To minimize excessive water loss, when the sun is shining brightly and photosynthesis is running at top speed, stomates are open. At night, though, most plants close their stomates.

Plants must keep their stomates open long enough to allow photosynthesis to take place but not so long that they lose too much water.

Guard cells are modified epithelial cells that control the opening and closing of the stomates in response to changes in water pressure. When guard cells absorb water by osmosis and become **turgid**, they curve like hot dogs, causing the stomate to open. When guard cells lose water and become flaccid, the stomate closes. Figure 10.6 is a sketch of open and closed guard cells and stomates in a leaf.

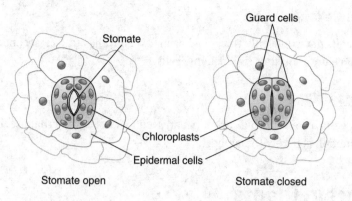

FIGURE 10.6 Stomates and guard cells

Types of Plant Tissue

Just as there are different cell and tissue types in animals, plants too have different cell and tissue types. Plants consist of three main tissue types: dermal, vascular, and ground tissue. A fourth tissue type, meristem tissue or growth tissue, is found only in the growing tips of shoots and roots.

Dermal Tissue

Dermal tissue is the outer protective covering of plants and usually consists of a single layer of epidermal cells. On leaves, epidermal cells are protected by the cuticle, which is made of the waxy molecule cutin. Some leaves are also covered with tiny, spikelike projections called **trichomes**, which also protect the leaf. For the most part, epidermal cells do not contain chloroplasts and cannot photosynthesize. An important exception is guard cells, which are modified epidermal cells that contain chloroplasts and can photosynthesize.

Vascular Tissue

Vascular tissue transports water and nutrients up and down the plant. There are two types: xylem and phloem. Xylem consists of **tracheids** and **vessel elements**. Phloem consists of **sieve tube elements** and **companion cells**.

Ground Tissue

Ground tissue makes up all plant tissue besides dermal and vascular tissue. It consists of three cell types: parenchyma, collenchyma, and sclerenchyma.

Parenchyma Cells

Parenchyma cells are the traditional-looking plant cells. They have primary cell walls that are thin and flexible, and they lack secondary cell walls. The cytoplasm contains one or two large vacuoles. When the cell is turgid (swollen) with water, these cells lend support to the plant. They are found in all parts of the plant. Some, like mesophyll cells, contain chloroplasts. Others, like epidermal cells, do not.

Collenchyma Cells

Collenchyma cells have unevenly thickened primary cell walls but lack secondary cell walls. The "strings" of celery consist of collenchyma cells.

Sclerenchyma Cells

Sclerenchyma cells have very thick primary and secondary cell walls that are fortified with lignin. Their function is purely for support.

Transport in Plants

Just like animals, plants need to transport water, nutrients, and gases. Unlike animals, plants do not have blood, arteries, or a heart to accomplish this. Instead, they have xylem and phloem.

Xylem

Xylem consists of two types of elongated cells: **tracheids** and **vessel elements**. The secondary cell walls of tracheids are hardened with **lignin** and function to support the plant as well as to transport nutrients and water. Xylem is what makes up the stuff we call wood.

Xylem carries water and nutrients from the soil up to the tallest leaves against gravity with *no expenditure of energy*. Instead, they are pulled up by a combination of two phenomena: **transpirational pull** and **cohesion tension**. Transpiration is the evaporation of water from leaves. Cohesion refers to the fact that water molecules are attracted to each other and stick together. The transpirational pull–cohesion tension theory states that *for each molecule of water that evaporates from a leaf by transpiration, another molecule of water is drawn in at the root to replace it*. The absorption of sunlight drives transpiration by causing water to evaporate from the leaf.

Several factors affect the rate of transpiration and loss of water from a leaf.

- High humidity slows down transpiration, while low humidity speeds it up.
- Wind can reduce humidity near the stomates and thereby increase transpiration.
- Increased light intensity increases photosynthesis, thereby increasing both the amount of water vapor to be transpired and the rate of transpiration.
- Closing stomates stop transpiration.

Phloem

Phloem vessels are made of chains of two types of cells: **sieve tube elements** and **companion cells**. They carry sugar from the photosynthetic leaves to the rest of the plant by a process called **translocation**. Sugar is stored in the roots. Unlike transport in the xylem, *this process requires energy*.

Plant Reproduction

Plants can reproduce both asexually and sexually.

Asexual Reproduction

Plants can clone themselves or reproduce asexually by **vegetative propagation**. In this process, a piece of the vegetative part of a plant—the root, stem, or leaf—produces an entirely new plant genetically identical to the parent plant. Examples are grafting, cuttings, bulbs, and runners.

Sexual Reproduction in Flowering Plants

The flower is the sexual organ of a plant. Figure 10.7 shows the structure and function of the parts of the flower.

1. **Petals.** Brightly colored, modified leaves found just inside the circle of sepals; attract animals that pollinate the plant

2. **Sepals.** Outermost circle of leaves; are green and closely resemble ordinary leaves; enclose the bud before it opens and protects the flower while it develops

3. **Pistils or carpels.** Female part of the flower; produce the female gametophytes; each consists of an ovary, a stigma, and a style

4. **Ovary.** Swollen part of pistil that contains the ovule, where one or more ova are produced by meiosis

5. **Ovule.** The structure within the ovary where the ova (female gametophytes) are produced

6. **Style.** Long, usually thin stalk of the pistil

7. **Stigma.** Sticky top of the style where pollen lands and germinates

8. **Stamen.** Male part of the flower, made up of anther and filament

9. **Anther.** Male part of the flower where sperm (pollen) are produced by meiosis

10. **Filament.** Threadlike structure that supports the anther

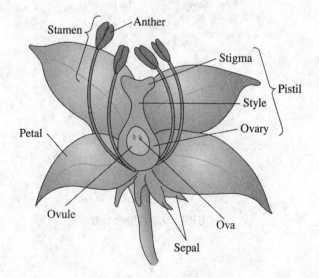

FIGURE 10.7 The flower

Pollination and Fertilization in Flowering Plants

Sexual reproduction begins with pollination. One pollen grain containing three haploid nuclei—one tube nucleus and two sperm nuclei—lands on the sticky stigma of the flower. The pollen grain absorbs moisture and germinates or sprouts, producing a pollen tube that burrows down the style into the ovary. The two sperm nuclei travel down the pollen tube into the ovary. Once inside the ovary, the two sperm nuclei enter the ovule through the micropyle. One sperm nucleus fertilizes the egg and becomes the embryo ($2n$). The other sperm nucleus fertilizes the two polar bodies and becomes the triploid ($3n$) **endosperm** or cotyledon, the food for the growing embryo.

This process is known as double fertilization because two fertilizations occur. After fertilization, the ovule becomes the **seed** and the ripened ovary becomes the **fruit**. In monocots, food reserves remain in the endosperm. In dicots, the food reserves of the endosperm are transported to the cotyledons, and consequently, the mature dicot seed lacks **endosperm**. In the monocot coconut, the endosperm is liquid.

Double fertilization can be seen as:

1. Sperm + Ovum → Embryo = $2n$
2. Sperm + 2 Polar bodies → Cotyledon (Food for the growing embryo) = $3n$

The Seed

The seed consists of a protective seed coat, an embryo, and the cotyledon or endosperm—food for the growing embryo. The embryo consists of the **hypocotyl**, **epicotyl**, and **radicle**. The **hypocotyl** becomes the lower part of the stem and the roots. The **epicotyl** becomes the upper part of the stem. The **radicle**, or embryonic root, is the first organ to emerge from the germinating seed.

Figure 10.8 shows a dicot seed (like a peanut) split in half. A monocot seed, like corn, does not split in half. In addition, the food source in a monocot is endosperm instead of cotyledon.

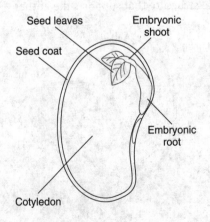

FIGURE 10.8 The seed

Alternation of Generations

The sexual life cycle of plants is characterized by the alternation of generations in which haploid (n) and diploid ($2n$) generations alternate with each other. The **gametophyte** (n) produces gametes by mitosis that fuse during fertilization to yield $2n$ zygotes. Each zygote develops into a **sporophyte** ($2n$) that produces haploid spores (n) by meiosis. Each haploid spore forms a new gametophyte, and the cycle continues, as seen in Figure 10.9.

> **REMEMBER**
> The gametophyte generation is haploid (n).
>
> The sporophyte generation is diploid ($2n$).
>
>

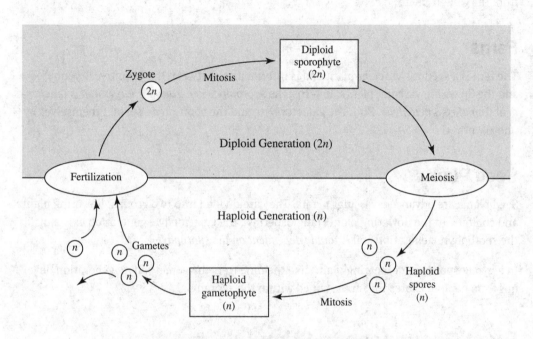

FIGURE 10.9 Alternation of generations

TABLE 10.3 Vocabulary for Alternation of Generations

Term	Definition
Antheridium	Structure that produces sperm, develops on the gametophyte
Archegonium	Structure that produces eggs, develops on the gametophyte
Gametophyte	A haploid adult plant
Megaspores	Produced by large female cones and develops into female gametophytes
Microspores	Produced by small male cones and develops into male gametophytes or pollen grains
Protonema	Branching filaments one-cell thick, produced by germinating moss spores, becomes the gametophyte in moss
Sporangia	Located on the tip of the mature sporophyte, where meiosis occurs, producing haploid spores
Sporophyte	A diploid adult plant
Sori	Raised spots located on the underside of sporophyte ferns, clusters of sporangia

Mosses and Other Bryophytes

Mosses are the green, carpetlike plants seen growing in damp forests, sometimes on fallen logs. They are primitive plants. The *gametophyte generation dominates* the life cycle. This means that the organism is haploid (*n*) for most of its life cycle, and the sporophyte (2*n*) is dependent on the gametophyte.

The gametophyte obtains nutrients by photosynthesis. The sporophyte obtains its nutrients from the gametophyte.

Ferns

The fern is a seedless vascular plant and is intermediate between the primitive bryophytes and the flowering vascular plants. In ferns, the *sporophyte generation is larger and is independent from the gametophyte*. Both the gametophyte and the sporophyte sustain themselves by photosynthesis.

Seed Plants

Seed plants are advanced, vascular plants. They are divided into two groups: flowering plants and conifers. In the flowering plants (angiosperms), the *gametophyte* generation exists inside the sporophyte generation and *is totally dependent on the sporophyte*.

In a gymnosperm (cone-bearing plant) like the pine tree, the gametophyte generation develops from haploid spores that are retained within the sporangia.

Plant Responses to Stimuli

Although plants cannot run away from stimuli, they can still react to stimuli.

Hormones

Plant hormones help coordinate growth, development, and response to environmental stimuli. They are produced in very small quantities. However, they have a profound effect on the plant because the hormone signal is amplified. A plant's response to a hormone usually depends not so much on absolute quantities of hormones but on relative amounts. Hormones can have multiple effects on a plant, and they can work synergistically with other hormones or in opposition to them. The following is an overview of plant hormones and what they stimulate.

Auxins

- Phototropisms occur due to an unequal distribution of **auxins**.
- Auxins enhance apical dominance, the preferential growth of a plant upward (toward the sun) rather than laterally. The terminal bud actually suppresses lateral growth by suppressing development of axial buds.
- Auxins stimulate stem elongation and growth by softening the cell wall.
- The first plant hormone discovered was auxin.

- Indoleacetic acid (IAA) is a naturally occurring auxin.
- A human-made auxin, 2,4-D, is used as a weed killer.
- Auxins are used as rooting powder to develop roots quickly in a plant cutting.
- A synthetic auxin sprayed on tomato plants will induce fruit production without pollination. This results in seedless tomatoes.

Cytokinins

- **Cytokinins** stimulate cytokinesis and cell division.
- They delay senescence (aging) by inhibiting protein breakdown. (Florists spray cut flowers with cytokinins to keep them fresh.)

Gibberellins

- **Gibberellins** promote stem and leaf elongation.
- They induce bolting, the rapid growth of a floral stalk. When a plant, such as broccoli, which normally grows close to the ground, enters the reproductive stage, it sends up a very tall shoot on which the flower and fruit develop. This is a mechanism to ensure pollination and seed dispersal.

Abscisic Acid (ABA)

- **Abscisic acid** inhibits growth and promotes seed dormancy.
- It enables plants to withstand drought.
- It closes stomates during times of water stress.

Ethylene

- This plant hormone is a gas.
- **Ethylene** gas promotes ripening, which in turn triggers increased production of ethylene gas. "One bad apple spoils the whole barrel." This is an example of positive feedback.
- Commercial fruit sellers pick perishable fruit before they are ripe, while still hard. When the fruit arrive at their destination, they are sprayed with ethylene gas to hasten the ripening. In contrast, apples are kept in an environment of CO_2 to eliminate exposure to ethylene gas and thus keep the apples from ripening or rotting. In this way, apples can be stored for long periods of time.

Tropisms

A **tropism** is the growth of a plant toward or away from a stimulus. Examples are thigmotropisms (touch), geotropisms or gravitropisms (gravity), and phototropisms (light). A growth of a plant toward a stimulus is known as a positive tropism, while a growth away from a stimulus is a negative tropism.

Phototropisms result from an unequal distribution of **auxins** that accumulate on the side of the plant away from the light. Since auxins cause growth, the cells on the shady side of the plant enlarge and the stem bends toward the light.

Geotropisms result from an interaction of auxins and statoliths, specialized plastids containing dense starch grains.

Animal Physiology

This chapter offers a review of nonhuman animals and will help you gain a broader understanding into adaptations common to all animals. The animals discussed are commonly studied in many biology courses.

Animals are multicellular eukaryotes. All are **heterotrophs** and acquire nutrients by **ingestion**. This chapter reviews general concepts in the areas of movement and locomotion, body temperature regulation, and excretion. This is followed by specific information about three representative animals in three different phyla: hydra (Cnidaria), earthworm (Annelida), and grasshopper (Arthropoda).

Movement and Locomotion

REMEMBER

An endoskeleton grows with the animal; an exoskeleton does not.

Movement is a characteristic of all animals. Movement can refer to the beating of cilia or the waving of tentacles to capture prey. It can also mean **locomotion**, movement from place to place. Some animals like hydra or sponges are **sessile**, meaning that they do not move. The hydra (phylum Cnidaria) feeds by moving its tentacles and stinging prey that swims near enough to touch.

Most animals spend their time and energy capturing food, seeking a mate, or escaping danger. Some mollusks, like clams, have a mantle, which secretes a shell that offers protection. Some arthropods, like crabs and grasshoppers, have an **exoskeleton** consisting mostly of the polysaccharide **chitin**, which *does not grow with the animal* and must be shed periodically. Not only does this exoskeleton protect the soft body inside, but in combination with muscles, it also enables the animal to move rapidly. Animals like the nematodes (roundworms), flatworms (planaria), and annelids (earthworms) have a **hydrostatic skeleton**, a closed body compartment filled with fluid. Muscles change the shape of this fluid-filled compartment, enabling the animal to move from place to place. Chordates, like frogs, cats, and humans, have an **endoskeleton** made of bone and cartilage that *grows as the animal grows*. Bones are connected to each other at joints by **ligaments**, while **tendons** connect bones to muscles.

Body Temperature Regulation

Most life exists only within a fairly narrow range, from 0°C, the temperature at which water freezes, to about 50°C. Animals must either seek out or create a suitable environment for themselves. The oceans are the most stable environment and experience the least fluctuation in environmental temperatures.

Temperatures on land, however, fluctuate enormously. Therefore, temperature regulation, like water conservation, became a problem for animals when they moved to land millions of years ago.

For example, the size of the ears in a jackrabbit can be correlated to the climate it lives in. Jackrabbits that evolved in cold, northern regions have small ears close to the head to minimize heat loss. Rabbits that evolved in warm, southern regions have long ears to dissipate heat from the many capillaries that make their ears appear pink.

Animals can also regulate body temperatures by changes in behavior. Here are some examples.

- A snake can warm itself in the sun and cool off by hiding in the shade.
- Animals on a cold prairie in winter huddle to decrease heat loss.
- Bees swarming in a hive raise the temperature inside the hive.
- Dogs pant and sweat through their tongues.
- Elephants lack sweat glands, but they wet down their thick skins with water and flap their ears, which are rich in capillaries.
- Humans shiver and jump around to keep warm.

Clarifying Terms

1. Cold-blooded and warm-blooded are not scientific terms.
2. **Ectotherm** means heated from outside and is probably closest in meaning to cold-blooded.
3. **Endotherm (homeotherm)** is the scientific word for warm-blooded. It means maintaining a constant body temperature despite fluctuations in the environmental temperature. Among animals, only birds and mammals are endotherms or homeotherms. Although being warm-blooded requires enormous energy, it may have given birds and mammals an edge in ancient Earth when the dominant animal was the reptile. Mammals and birds can be active at any time, while reptiles can be active only when the temperature permits it.

Excretion

Excretion is the removal of metabolic wastes. These include water, carbon dioxide, and nitrogenous wastes. There are three different types of nitrogenous wastes.

Ammonia

- Very soluble in water and highly toxic
- Excreted generally by organisms that live in water, including hydra and fish

Urea

- Not as toxic as ammonia
- Excreted by earthworms and humans
- In mammals, is formed in the liver from ammonia

Uric Acid

- Pastelike substance that is not soluble in water and therefore not very toxic
- Excreted by insects, many reptiles, and birds, with a minimum of water loss

Different excretory mechanisms have evolved in various organisms for the purpose of removing metabolic wastes. Table 11.1 lists some organisms matched with their structures.

TABLE 11.1 Excretion in Various Animals

Organism	Structure	Nitrogenous Waste
Hydra	None	Ammonia
Platyhelminthes (planaria)	Flame cells	Ammonia
Earthworms	Nephridia (metanephridia)	Urea
Insects	Malpighian tubules	Uric acid
Humans	Nephrons	Urea

Hydra—Phylum Cnidaria

Nutrition

In cnidarians like hydra and jellyfish, digestion occurs in the **gastrovascular cavity,** which has only one opening, the mouth. The animal has a two-way digestive tract, which means that food enters the same opening as waste exits. Cells of the gastrodermis (lining of the gastrovascular cavity or gastrocoel) secrete digestive enzymes into the cavity to aid in extracellular digestion where the main part of digestion occurs. Since the cnidarians are animals, their cells contain **lysosomes** that carry out intracellular digestion as well as extracellular digestion. See Figure 11.1.

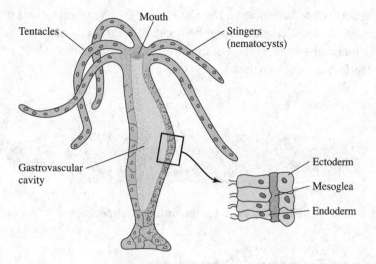

FIGURE 11.1 The hydra

Body Plan and Symmetry

The basic body plan of the hydra is a **polyp**, while the body plan of the jellyfish is the **medusa**. The symmetry of all animals in this phylum is primitive and radial. The animal has only two cell layers, the **ectoderm** and **endoderm**. The layers are held together by a middle layer called the **mesoglea** (middle glue). Every cell is in direct contact with its environment, and therefore the hydra has no need of a circulatory system.

Nervous System

All cnidarians have unique cells called **cnidocytes** that contain stingers, called **nematocysts**. Response to the environment is controlled by a primitive nervous system, a nerve net, where impulses travel in all directions from any site. As a result, the entire animal responds to a single stimulus. For example, if you knock the dish a hydra is in, its entire body will respond by temporarily shrinking into a tiny ball.

Reproduction

Cnidarians reproduce sexually, as well as asexually, by **budding**. A bud is a genetically identical but miniature version of the parent that forms within or on the parent. Ultimately, it breaks free to become an independent being.

Earthworm—Phylum Annelida

Nutrition

The earthworm plays an important ecological role as it burrows in the ground and creates tunnels that aerate the soil. The digestive tract of the earthworm is a long, straight tube. The mouth ingests decaying organic matter along with soil. From the mouth, food moves to the esophagus and then to the **crop**, where it is stored. Posterior to the crop, the **gizzard**, which consists of thick muscular walls, grinds up the food with the help of sand and soil, which were ingested along with the organic matter. The rest of the digestive tract consists of the intestines where chemical digestion and absorption occur. Absorption is enhanced by the presence of a large fold in the upper surface of the intestine, called the typhlosole, which greatly increases the surface area. See Figure 11.2.

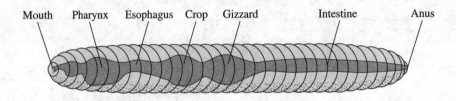

Mouth   Pharynx   Esophagus   Crop   Gizzard   Intestine   Anus

FIGURE 11.2 Earthworm digestive tract

Nervous and Transport Systems

In the earthworm, the exchange of respiratory gases—oxygen and carbon dioxide—between the environment and cells occurs passively by **diffusion** through moist skin. Earthworms are said to have an **external respiratory surface** because diffusion of these gases occurs at the animal's surface. The heart consists of five pairs of **aortic arches** that pump blood through the body in arteries, veins, and capillaries. Since blood never normally leaves these blood vessels, we say the earthworm has a **closed circulatory system**. Oxygen is carried by hemoglobin dissolved in red blood.

The brain of the earthworm consists of two dorsal, solid, fused ganglia that connect to a solid, ventral nerve cord.

Excretion

The earthworm has paired **nephridia** in every body segment to remove the nitrogenous waste urea.

Reproduction

The earthworm is a **hermaphrodite**, meaning it has both male and female sex organs.

Grasshopper—Phylum Arthropoda

Nutrition

Like the earthworm, the grasshopper has a digestive tract that consists of a long tube that also contains a **crop** and **gizzard**. However, there are several differences between the two animals. The grasshopper has specialized mouthparts for tasting, biting, and crushing food and a gizzard that contains plates made of chitin that help grind food. In addition, the digestive tract is also responsible for removing the nitrogenous waste uric acid from the animal. **Malpighian tubules** serve this function. See Figure 11.3.

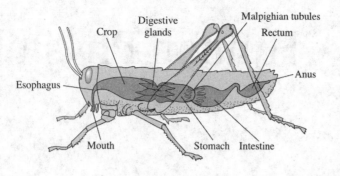

FIGURE 11.3 Grasshopper

Nervous and Transport Systems

The nervous system of the grasshopper is similar to that of the earthworm, but the transport system is different. The grasshopper heart is tubular, and the animal lacks capillaries. The grasshopper has an **open circulatory system** where blood normally leaves the artery and moves through interconnected **sinuses** or **hemocoels**, which are spaces surrounding the organs. Arthropod blood does not carry either hemoglobin or oxygen.

Exchange of Respiratory Gases

The grasshopper and other arthropods and crustaceans have an **internal respiratory surface** because exchange of oxygen and carbon dioxide occurs inside the animal. Air enters the body through **spiracles** and travels through a system of **tracheal tubes** into the **hemocoels** or **sinuses**, where diffusion occurs. In arthropods and in some mollusks, oxygen is carried by hemocyanin, a molecule similar to hemoglobin but with copper, instead of iron, as its core atom.

Human Physiology

This chapter includes a review of the following topics about human physiology—digestion, gas exchange, circulation, endocrine system, nervous system, excretion, and muscles. Two remaining topics in human physiology, human reproduction and the immune system, are included in the next two chapters.

Digestion

The human digestive system has two important functions: breaking down large food molecules into smaller, usable molecules and absorbing these smaller molecules. Fats get broken down into glycerol and fatty acids, starch into monosaccharides, nucleic acids into nucleotides, and proteins into amino acids. Vitamins and minerals are small enough to be absorbed without being digested.

The digestive tract is about 30 feet long and is made of smooth (involuntary) muscle that pushes the food along the digestive tract by a process called peristalsis. The muscles of the digestive tract are controlled by the autonomic nervous system. Figure 12.1 is a chart detailing the structures and function of the human digestive system.

Mouth

- Mechanical and chemical digestion begins here.

- The enzyme **salivary amylase** in saliva begins starch digestion.

- The tongue and differently shaped teeth work together to break down food mechanically.

- The type of teeth an animal has is a reflection of its dietary habits. Humans are omnivores and have three different types of teeth: incisors for cutting, canines for tearing, and molars for grinding.

Esophagus

- No digestion occurs here.

- After swallowing, food is directed into the esophagus and away from the windpipe by the **epiglottis**, a flap of cartilage in the back of the **pharynx** (throat).

- The esophagus transports food from the throat to the stomach.

REMEMBER
Starch digestion begins in the mouth.

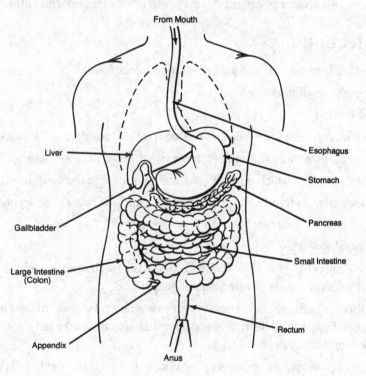

FIGURE 12.1 Human digestive system

Stomach

- Both mechanical and chemical digestion occur here.
- Protein digestion begins in the stomach.
- The stomach's thick, muscular wall churns food mechanically and secretes gastric juice, which contains hydrochloric acid and enzymes that digest proteins.
- Hydrochloric acid begins the breakdown of muscle (meat) and activates the inactive enzyme **pepsinogen** to become pepsin, which digests protein.
- The enzyme rennin aids in the digestion of the protein in milk.
- The pH in the stomach is acidic, ranging from 2 to 3.
- The cardiac sphincter at the top of the stomach keeps acidified food in the stomach from backing up into the esophagus and burning it. The pyloric sphincter at the bottom of the stomach keeps the food in the stomach long enough to be digested.
- Excessive acid can cause an ulcer to form in the esophagus, the stomach, or the duodenum (the upper intestine). We now know that a common cause of ulcers is a particular bacterium, *Helicobacter pylori*, which can be effectively treated with antibiotics.

Small Intestine

- All digestion is completed and nutrients are absorbed here.
- The pH in the small intestine is 8.
- It is 6 meters long.
- All digestion is completed in the duodenum, the first 10 inches of the small intestine.
- The intestinal enzymes are amylases, proteases, lipases, and nucleases.
- Pancreatic amylases, which digest starch, are secreted into the small intestine.
- **Peptidases**, such as trypsin and chymotrypsin, continue to break down proteins.
- Nucleases hydrolyze nucleic acids into nucleotides.
- Lipases break down fats.
- Millions of fingerlike projections called villi line the small intestine and absorb all nutrients that were previously released from digested food.
- Each **villus** contains capillaries, which absorb amino acids, vitamins, and monosaccharides directly into the bloodstream, and a **lacteal**, which absorbs fatty acids and glycerol into the lymphatic system.
- Villi have microscopic appendages called microvilli that further enhance the rate of absorption. Figure 12.2 is a drawing of a villus.

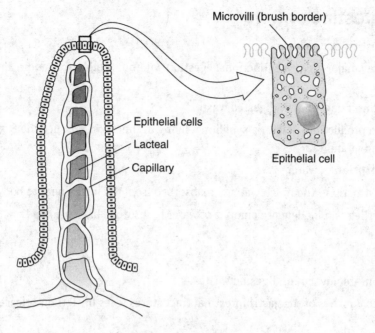

Microvilli (brush border)

Epithelial cells

Lacteal

Capillary

Epithelial cell

FIGURE 12.2 Villus

Liver

- Produces **bile** that emulsifies fats
- Bile = pH 11; neutralizes chyme (acidified food from stomach) entering small intestine
- Sends bile to the gallbladder until its release into the small intestine
- Has other functions besides digestion
  1. Breaks down and recycles red blood cells
  2. Detoxifies blood—removes alcohol and drugs
  3. Produces cholesterol necessary for structure of cell membranes
  4. Produces the nitrogenous waste urea from protein metabolism

REMEMBER
Bile is not an enzyme, but it breaks down fats.

Gallbladder

- Stores bile that is produced in liver
- Bile emulsifies fats in small intestine
- Body can function well without a gallbladder

Pancreas

The pancreas is a gland that serves two important functions in the digestive system.

- Produces enzymes that break down carbohydrates (amylases), proteins (peptidases), lipids (lipases), and nucleic acids, and secretes them into the small intestine
- Produces **sodium bicarbonate**, a base that neutralizes stomach acid, enabling intestinal enzymes, which require a basic environment, to be effective

As part of the endocrine system, it produces hormones to control blood sugar levels.

Large Intestine or Colon

- No digestion occurs here
- Has three major digestive functions: egestion, vitamin production, and reabsorption of water
- **Egestion**—removal of undigested waste
- Vitamin production—bacteria symbionts living in the colon produce the B vitamins, folic acid, and vitamin K
- **Reabsorption** of water—
  1. Constipation—too much water is reabsorbed from the intestine into body
  2. Diarrhea—an inadequate amount of water is absorbed back into body

Rectum

- Egestion—removal of undigested waste
- Last 7 to 8 inches of the gastrointestinal tract stores feces until their release through the anus

Gas Exchange

In humans, air enters the nasal cavity and is moistened, warmed, and filtered. From there, air passes through the larynx and down the trachea and bronchi into the tiniest bronchioles, which end in microscopically tiny air sacs called **alveoli**. Here is where diffusion of respiratory gases occurs. See Figure 12.3.

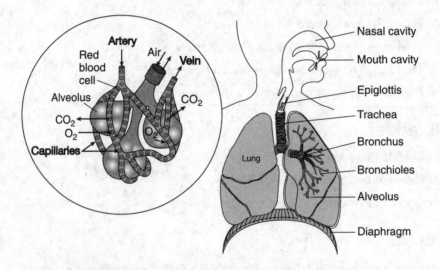

FIGURE 12.3 Adaptations for human respiration

Humans have an internal respiratory surface because respiratory gases are exchanged deep inside the body. The rib cage expands and forces the diaphragm to contract and move downward, thus expanding the chest cavity and decreasing the internal pressure. Air is drawn into the lungs by *negative pressure* because the internal pressure inside the chest cavity is lower than the air pressure surrounding the body.

The **medulla** in the brain sets the breathing rhythm by monitoring carbon dioxide levels in the blood and by sensing changes in the pH of the blood. A blood pH lower than 7.4 triggers autonomic nerves from the medulla to increase the breathing rate to rid the body of more carbon dioxide. The concentration of oxygen in the blood usually has little effect on the breathing control centers.

Transport of Oxygen and Carbon Dioxide

Oxygen is carried in the human blood by the respiratory pigment **hemoglobin**, which combines loosely with oxygen molecules to form the molecule **oxyhemoglobin**.

Carbon dioxide, the by-product of cell respiration, is released from every cell and dissolves in the blood. Carbon dioxide is carried in the plasma as part of the reversible blood-buffering **carbonic acid–bicarbonate ion system**, which maintains the blood at a constant pH of 7.4. Very little carbon dioxide is transported by hemoglobin.

Circulation

Human circulation consists of a closed circulatory system with arteries, veins, and capillaries. Table 12.1 shows the characteristics of each structure.

TABLE 12.1 Structure and Function of Blood Vessels

Vessel	Function	Structure
Artery	Carries blood away from the heart under enormous pressure	Walls made of thick layer of elastic, smooth muscle. Can withstand high pressure and can contract and expand as needed.
Vein	Carries blood back to the heart under very little pressure	Walls do not contain thick layer of muscle. Has valves to help prevent backflow. Located within skeletal muscle, which propels blood upward and back to heart as the body moves and muscles contract.
Capillary	Allows for diffusion of nutrients and wastes between cells and blood	Walls are one-cell thick and so small that blood cells travel only single file. Blood travels slowly here to allow time for diffusion of nutrients and wastes.

Blood

Blood consists of several different cell types suspended in a liquid matrix called **plasma**. The average human body contains 4 to 6 liters of blood. Table 12.2 is a chart that gives basic information about blood.

The Mechanism of Blood Clotting

Blood clotting is a complex mechanism that begins with the release of clotting factors from platelets and damaged tissue. It involves a complex set of reactions. Anticlotting factors constantly circulate in the plasma to prevent the formation of a clot or thrombus, which can cause serious damage in the absence of injury. **Serum** is plasma minus clotting factors. Calcium is necessary for normal blood clotting.

Here is the pathway of normal clot formation:

Damaged tissue and platelets release

↓

Thromboplastin + Ca^{++}

↓ Stimulates

Prothrombin → **Thrombin**

(Inactive)   (Active)

↓ Stimulates

Fibrinogen → **Fibrin** (Clot)

(Inactive)   (Active)

TABLE 12.2 Components of Blood

Component	Scientific Name	Properties
Plasma	(None)	Liquid portion of the blood.
		Contains clotting factors, hormones, antibodies, dissolved gases, nutrients, and wastes.
		90% water.
Red blood cells	Erythrocytes	Carry hemoglobin and oxygen.
		Do not have a nucleus.
		Circulate about 120 days.
		Formed in the bone marrow and recycled in the liver.
White blood cells	Leukocytes	Fight infection.
		Formed in the bone marrow.
		Die fighting infection and are one component of pus.
Platelets	Thrombocytes	Clot blood.
		Cell fragments that are formed in the bone marrow from megakaryocytes.

The Heart

The heart is located beneath the sternum and is about the size of your clenched fist. Figure 12.4 is a diagram of the heart. It beats about 70 beats per minute and pumps about 5 liters of blood—the total volume of blood in the body—each minute. Two atria (atria, plural; atrium, singular) receive blood from the cells of the body, and two ventricles pump blood out of the heart.

The heart itself has its own pacemaker, the **sinoatrial (SA) node**, which sets the timing of the contractions of the heart. Electrical impulses travel through the cardiac and body tissues to the skin, where they can be detected by an electrocardiogram (EKG). The heart's pacemaker is influenced by a variety of factors: the nervous system, hormones such as adrenaline, and body temperature.

Blood pressure is lowest in the veins and highest in the arteries when the ventricles contract. The average blood pressure for all normal resting adults is 120/80. The **systolic** number (120) is a measurement of the pressure when the ventricles contract, while the **diastolic** number (80) is a measure of the pressure when the heart relaxes. Remember that the right side of the heart in the figure is actually the left side of the heart in the body.

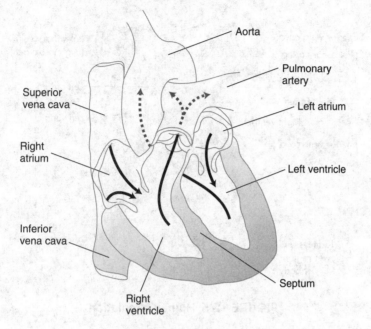

FIGURE 12.4 The heart

Pathway of Blood

Blood enters the heart through the vena cava. From there it continues to the:

1. Right atrium
2. Right atrioventricular (AV) valve or tricuspid valve
3. Right ventricle
4. Pulmonary semilunar valve
5. Pulmonary artery
6. Lungs
7. Pulmonary vein
8. Left atrium
9. Left atrioventricular (AV) valve or bicuspid valve
10. Left ventricle
11. Aortic semilunar valve
12. Aorta
13. To all the cells in the body
14. Returns to the heart through the vena cava

Blood circulates through the coronary circulation (heart), the renal circulation (kidneys), and the hepatic circulation (liver). The pulmonary circulation includes the pulmonary artery, the lungs, and the pulmonary vein.

One thing to remember is that the pulmonary artery is the only artery that carries deoxygenated blood and the pulmonary vein is the only vein that carries oxygenated blood. Figure 12.5 is a drawing of the human circulatory system.

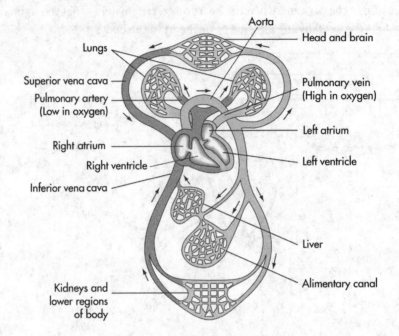

FIGURE 12.5 Human circulation

Endocrine System

Animals have two major regulatory systems that release chemicals: the endocrine system and the nervous system. Even though the two systems are separate, they work together to regulate the body, to maintain **homeostasis**. The endocrine system secretes **hormones**, while the nervous system secretes neurotransmitters. In one case, epinephrine functions both as the fight-or-flight hormone secreted by the adrenal gland and as a neurotransmitter that sends a message from one neuron to another.

Hormones are produced in **ductless** (endocrine) **glands** and move through the blood to a specific target cell, tissue, or organ that can be far from the original endocrine gland. Hormones can produce either an immediate short-lived response or a dramatic long-term development of an entire organism. An example of a short-lived response is the way adrenaline (epinephrine) causes the flight-or-fight response. An example of a long-term response is the way ecdysone controls metamorphosis in insects.

Tropic hormones are hormones that stimulate other glands to release hormones and can have a far-reaching effect. For example, the anterior pituitary in the brain releases thyroid-stimulating hormone (TSH), which stimulates the thyroid in the neck region to release thyroxin. Other types of chemical messengers reach their target by special means. **Pheromones** in the urine of a dog carry a message between different individuals of the same species. In vertebrates, nitric oxide (NO), a gas, is produced by one cell and diffuses to and affects only neighboring cells before it is broken down.

The Hypothalamus

The **hypothalamus**, located in the brain, plays a special role in the body; it is the *bridge between the endocrine and nervous systems*. The hypothalamus acts as part of the nervous system when, in times of stress, it sends electrical signals to the adrenal gland to release adrenaline. It acts as an endocrine gland when it produces oxytocin and antidiuretic hormone that it stores in the posterior pituitary. The hypothalamus also contains the body's thermostat and centers for regulating hunger and thirst. Table 12.3 is an overview of the other glands and hormones of the endocrine system.

TABLE 12.3 Endocrine Hormones

Gland	Hormone	Effect
Anterior pituitary	• Growth hormone (GH)	Stimulates growth of bones
	• Luteinizing hormone (LH)	Stimulates ovaries and testes
	• Thyroid-stimulating hormone (TSH)	Stimulates thyroid gland
	• Adrenocorticotropic hormone (ACTH)	Stimulates adrenal cortex to secrete glucocorticoids
	• Follicle-stimulating hormone (FSH)	Stimulates gonads to produce sperm and ova
Posterior pituitary	• Oxytocin	Stimulates contractions of uterus and mammary glands
	• Antidiuretic hormone (ADH)	Promotes retention of water by kidneys
Thyroid	• Thyroxin	Controls metabolic rate
	• Calcitonin	Lowers blood calcium levels
Parathyroid	Parathormone	Raises blood calcium levels
Adrenal cortex	Glucocorticoids	Raises blood sugar levels
Adrenal medulla	• Epinephrine (adrenaline)	Raises blood sugar levels by increasing rate of glycogen breakdown by liver
	• Norepinephrine (noradrenaline)	
Pancreas—islets of Langerhans	• Insulin	Lowers blood glucose levels
	• Glucagon	Raises blood glucose levels
Thymus (in neck)	Thymosin	Stimulates T lymphocytes as part of the immune response
Pineal (in brain)	Melatonin	Involved in biorhythms
Ovaries	• Estrogen	Stimulate uterine lining, promote development and maintenance of primary and secondary characteristics of female
	• Progesterone	Promote uterine lining growth
Testes	Androgens	Support sperm production and promote secondary sex characteristics

How Hormones Trigger a Response in Target Cells

There are two types of hormones, steroid and nonsteroid or polypeptide hormones. They stimulate target cells in different ways. These are illustrated in Figure 12.6.

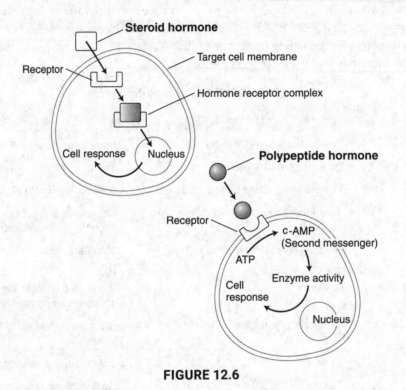

FIGURE 12.6

Lipids or steroid hormones diffuse directly through the plasma membrane and bind to a receptor inside the cell that triggers the cell's response.

Protein or polypeptide hormones (nonsteroid) cannot dissolve in the plasma membrane, so they bind to a receptor on the surface of the cell. Once the hormone (the first messenger) binds to a receptor on the surface of the cell, it triggers a secondary messenger such as c-AMP, which converts the extracellular chemical signal to a specific response inside the cell.

Feedback Mechanisms

A feedback mechanism is a self-regulating mechanism that increases or decreases an action or the level of a particular substance.

Positive Feedback

Positive feedback enhances an already existing response. During childbirth, for example, the pressure of the baby's head against sensors near the opening of the uterus stimulates more uterine contractions, which causes increased pressure against the uterine opening, which causes yet more contractions. This positive feedback loop brings labor to an end and the birth of a baby. This is very different from negative feedback.

Negative Feedback

Negative feedback is a common mechanism in the endocrine system (and elsewhere) that maintains homeostasis. A good example is how the body maintains proper levels of thyroxin. When the level of thyroxin in the blood is too low, the hypothalamus stimulates the anterior pituitary to release a hormone, thyroid-stimulating hormone (TSH), which stimulates the thyroid to release more thyroxin. When the level of thyroxin is adequate, the hypothalamus stops stimulating the pituitary.

Nervous System

The vertebrate nervous system consists of central and peripheral components.

- The central nervous system (CNS) consists of the brain and spinal cord.
- The peripheral nervous system (PNS) consists of all nerves outside the CNS.

The peripheral nervous system is then further divided and subdivided into various systems. The following is an overview of the peripheral nervous system.

Outline of the Peripheral Nervous System

Sensory—conveys information from sensory receptors or nerve endings
Motor—stimulates voluntary and involuntary muscles and consists of two systems:

| **Somatic System** | Controls the voluntary muscles |
| **Autonomic System** | Controls involuntary muscles |

Sympathetic

- Fight-or-flight response
- Increases heart and breathing rate
- Liver converts glycogen to glucose
- Bronchi of lungs dilate and increase gas exchange
- Adrenaline raises blood glucose levels

Parasympathetic

- Opposes the sympathetic system
- Calms the body
- Decreases heart/breathing rate
- Enhances digestion

The Neuron

The neuron is the basic functional unit of the nervous system and is illustrated in Figure 12.7. It consists of a cell body, which contains the nucleus and other organelles, and two types of cytoplasmic extensions, called dendrites and axons.

- Dendrites are sensory. They receive incoming messages from other cells and carry the electrical signal to the cell body. A neuron can have hundreds of dendrites.

- Axons transmit an impulse from the cell body outward to another cell. A neuron has only one axon, which can be several feet long in large mammals. Most axons are wrapped in a myelin sheath that protects the axon and speeds the impulse.

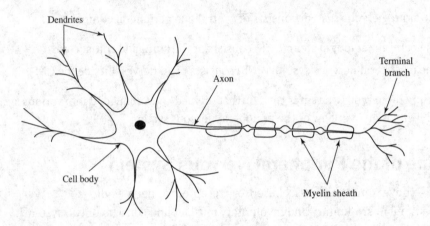

FIGURE 12.7 The neuron

The Reflex Arc

The simplest nerve response is a reflex arc. It is inborn, automatic, and protective. An example is the knee-jerk reflex, which consists of only a sensory and a motor neuron. You have experienced this during a physical exam. The doctor taps your knee with a hammer, and your foot kicks up involuntarily. The impulse moves from the sensory neuron in your knee to the motor neuron that directs the thigh muscle to contract. The spinal cord is not involved in this type of reflex.

A more complex reflex arc consists of three neurons: a sensory, a motor, and an interneuron or association neuron. This is illustrated in Figure 12.8. A sensory neuron transmits an impulse to the interneuron in the spinal cord that sends one impulse to the brain for processing and also one to the motor neuron to effect change immediately (at the muscle). This is the type of response that quickly jerks your hand away from a hot iron before your brain has figured out what occurred.

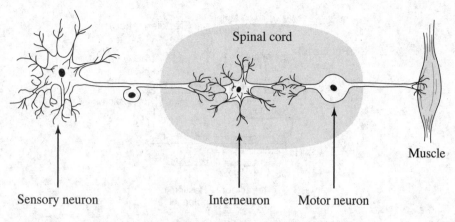

Spinal cord

Muscle

Sensory neuron

Interneuron   Motor neuron

FIGURE 12.8 The reflex arc

How a Neuron Functions

All living cells exhibit a membrane potential, a difference in electrical charge between the cytoplasm (negative charge) and extracellular fluid (positive charge). Physiologists measure this difference in membrane potential to be between -50 mV to -100 mV. The negative sign indicates that the inside of the cell is negative relative to the outside of the cell.

Resting Potential

A neuron at rest or unstimulated (resting potential) is **polarized** and has a membrane potential of about -70 mV. The sodium-potassium pump maintains this polarization by actively pumping ions out of the cell that leak inward. In order for the nerve to fire, a stimulus must be strong enough to overcome the resting threshold or resting potential. *The larger the membrane potential, the stronger the stimulus must be to overcome the resting potential and cause the nerve to fire.*

Action Potential

An **action potential**, or impulse, can be generated only in the axon of a neuron. When an axon is stimulated sufficiently to overcome the threshold, the permeability of a region of the membrane suddenly changes. Sodium channels open and sodium floods into the cell, down the concentration gradient. In response, potassium channels open and potassium floods out of the cell. This rapid movement of ions or **wave of depolarization** reverses the polarity of the membrane and is called an action potential. The action potential is localized and lasts a very short time.

The sodium-potassium pump restores the membrane to its original polarized condition by pumping sodium and potassium ions back to their original positions. This period of **repolarization**, which lasts a few milliseconds, is called the **refractory period**, during which the neuron cannot respond to another stimulus. The refractory period ensures that an impulse moves along an axon in one direction only since the impulse can move only to a region where the membrane is polarized. Figure 12.9 shows the axon of a neuron as an impulse passes from left to right, depolarizing the membrane in front of it.

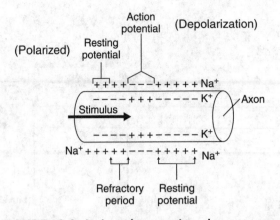

FIGURE 12.9 An impulse passing along an axon

The action potential is like a row of dominoes falling in order after the first one is knocked over. The first action potential generates a second action potential, which generates a third, and so on. The impulse moves along the axon, propagating itself *without losing any strength*. If the axon is myelinated, the impulse travels faster.

The action potential is an all-or-nothing event. Either the stimulus is strong enough to cause an action potential, or it is not. The body distinguishes between a strong stimulus and a weak one by the frequency of action potentials. A strong stimulus sets up more action potentials than a weak one does.

Figure 12.10 traces the events of a membrane undergoing sufficient stimulation to undergo an action potential.

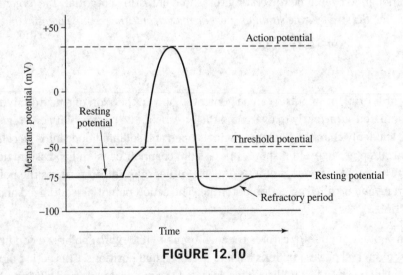

FIGURE 12.10

The Synapse

Although an impulse travels along an axon electrically, it crosses a synapse chemically. The cytoplasm at the terminal branch of the neuron contains many **vesicles**, each containing thousands of molecules of a **neurotransmitter**. Depolarization of the presynaptic membrane causes Ca^{++} ions to rush into the terminal branch through calcium-gated channels. This sudden rise in Ca^{++} levels stimulates the vesicles to fuse with the presynaptic membrane and release the neurotransmitter by exocytosis into the synapse, which sets up another action potential on the adjacent cell. Shortly after the neurotransmitter is released into the synapse, it is destroyed by an enzyme that stops the impulse at that point. The most common neurotransmitters are acetylcholine, serotonin, epinephrine, norepinephrine, dopamine, and GABA. In addition, many cells release the gas nitric oxide (NO) to stimulate other cells.

Figure 12.11 shows the terminal branch of the neuronal axon and the synapse.

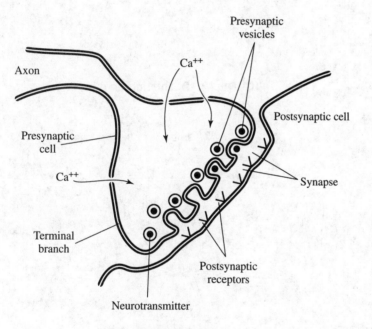

FIGURE 12.11

The Eye and the Ear

This section lists the structures and functions of the eye and the ear, and Figures 12.12 and 12.13 are illustrations.

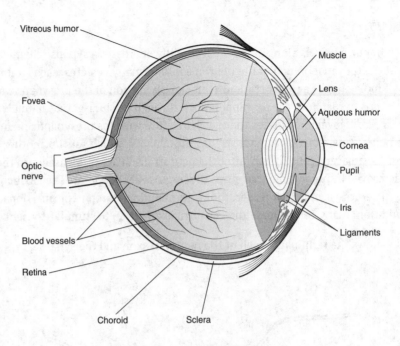

FIGURE 12.12 The human eye

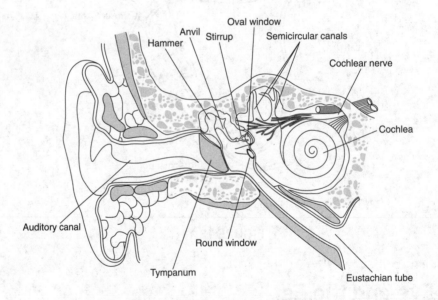

FIGURE 12.13 The human ear

Eye

- **Cones**—photoreceptors in the retina that distinguish different colors
- **Cornea**—tough, clear covering that protects the eye and allows light to pass through
- **Humor**—fluids that maintain the shape of the eyeball
- **Iris**—colored part of the eye that controls how much light enters the eye
- **Lens**—focuses light onto the retina
- **Pupil**—small opening in the middle of the iris
- **Retina**—converts light into nerve impulses that are carried to the brain
- **Rods**—photoreceptors in the retina that are extremely sensitive but do not distinguish different colors

Ear

- **Auditory canal**—ear canal, where sound enters
- **Cochlea**—fluid-filled part of inner ear, sends nerve impulses to brain
- **Ear bones**—hammer, anvil, and stirrup; transmit vibrations from eardrum to oval window
- **Eustachian tube**—equalizes pressure between environment and inner ear
- **Oval window**—sends waves of pressure to the cochlea
- **Semicircular canals**—fluid filled, helps you maintain your balance
- **Tympanum**—eardrum, vibrates as sound waves hit it

Excretion

Excretion is the removal of metabolic wastes. This includes removing carbon dioxide and water from cell respiration and nitrogenous wastes from protein metabolism. The organs of excretion in humans are the skin, lungs, liver, and kidneys.

Skin excretes sweat consisting of water and salts, including urea. Lungs excrete water vapor and carbon dioxide from the citric acid cycle. The liver does not excrete any substances from the body, but it is the site of deamination of amino acids and the production of urea. The kidneys excrete excess water and urea.

The Human Kidney

The kidney adjusts both the volume and the concentration of urine depending on the intake of water and salt and the production of urea. Humans have two kidneys supplied by blood from the **renal artery**. The kidneys filter about 1,500 liters of blood per day and produce, on average, 1.5 liters of urine. Like all terrestrial animals, we need to conserve as much water as possible, but we must balance the need to conserve water against the need to rid the body of soluble poisons.

> **TIP**
>
> The organs of excretion are:
> - skin
> - lungs
> - kidney
> - liver

The kidney is able to respond quickly to the changing requirements of the body because it is under hormonal control. A major hormone, **antidiuretic hormone (ADH)**, is released by the **posterior pituitary** and targets the **collecting tube** of the **nephron**. ADH regulates blood pressure by controlling how much water is reabsorbed by the kidneys.

The Nephron

The nephron, as shown in Figure 12.14, is the basic functional unit of the kidney. It consists of a cluster of capillaries, known as the **glomerulus**—which sits inside a cuplike structure called **Bowman's capsule**, a long narrow tube called the **tubule**, and the **loop of Henle**. Each human kidney contains about 1 million nephrons. The nephron carries out its job in four steps: **filtration**, **secretion**, **reabsorption**, and **excretion**.

Filtration

Filtration occurs by *diffusion*. It is *passive* and *nonselective*. The filtrate contains everything small enough to diffuse out of the glomerulus and into Bowman's capsule, including glucose, salts, vitamins, wastes such as urea, and other small molecules. From Bowman's capsule, the filtrate travels into the loop of Henle and then the collecting duct or tubule. From the collecting tubule, the filtrate trickles into the ureter and the urinary bladder for temporary storage and then to the **urethra** and out of the body.

Secretion

Secretion is the *active*, *selective* uptake of molecules that did not get filtered into Bowman's capsule. This occurs in the tubules of the nephron.

Reabsorption

Reabsorption is the process by which most of the water and solutes (glucose, amino acids, and vitamins) that initially entered the tubule during filtration are transported back into the capillaries and thus back to the body. This process occurs in the tubule, the loop of Henle, and the collecting tubule. The longer the loop of Henle, the greater is the reabsorption of water.

Excretion

Excretion is the removal of metabolic wastes, for example, nitrogenous wastes. Everything that passes into the collecting tubule is excreted from the body.

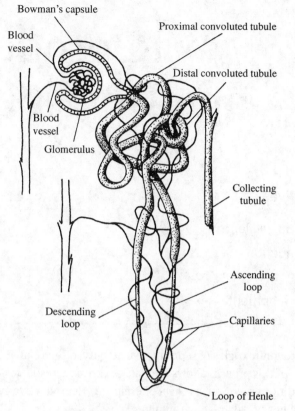

Bowman's capsule
Blood vessel
Proximal convoluted tubule
Distal convoluted tubule
Blood vessel
Glomerulus
Collecting tubule
Ascending loop
Descending loop
Capillaries
Loop of Henle

FIGURE 12.14 The nephron

Muscles

There are three types of muscle: smooth, cardiac, and skeletal.

- **Smooth** or **involuntary muscle** makes up the walls of blood vessels and the digestive tract. It does not have a striated appearance, hence the name. It is under the control of the autonomic nervous system.

- **Skeletal** or **voluntary (striated) muscles** are very large and multinucleate. They work in pairs; one muscle contracts while the other relaxes. The biceps and triceps in your upper arm are one such pair.

- **Cardiac muscle** is found in the heart and is striated like a skeletal muscle. It generates its own action potential. Individual heart cells will beat on their own, even when removed from the body.

The Sliding Filament Theory

Within the cytoplasm of each skeletal muscle cell are thousands of fibers called myofibrils that run parallel to the length of the cell. Myofibrils consist of thick and thin filaments. Each thin filament consists of **actin** proteins; each thick filament is composed of **myosin** proteins. Muscles contract as thick and thin filaments slide over each other.

CHAPTER 13

Reproduction and Development

Most animals show definite cycles of reproductive activity, often related to changing seasons. The periodic nature of this process allows animals to conserve resources and reproduce when environmental conditions favor the survival of offspring. Reproductive cycles are controlled by a combination of hormonal and environmental cues.

Animals may reproduce only asexually, only sexually, or alternate between the two. Asexual reproduction results in offspring that are genetically identical to the parents. This is an advantage when environmental conditions are stable.

Some eggs can develop by **parthenogenesis**, a process in which the egg develops without being fertilized and the adults that result are haploid. This is characteristic of honeybees, where haploid individuals are male drones and diploid individuals are female workers. Some sessile animals are **hermaphrodites** and can mate with any animal of their species. Both animals act as male and female, and both donate and receive sperm.

REMEMBER
Haploid is currently the term used exclusively to mean the n number of chromosomes.

Sexual reproduction offers increased variation among offspring and the possibility of greater reproductive success in a changing environment. During sexual reproduction, a small flagellated haploid sperm (n) fertilizes a larger, nonmotile haploid egg (n) to form a diploid ($2n$) zygote. The zygote then undergoes **cleavage**, **gastrulation**, and **organogenesis**. Sexually reproducing fish and amphibians carry out external fertilization, where the female sheds thousands of eggs to be fertilized by sperm directly in the environment. The likelihood that sperm and egg will actually fuse is low, and the rate of predation of those that actually form a zygote is high. To compensate, millions of eggs and sperm are released at one time. Fish and amphibians reproduce this way. Birds, reptiles, and mammals carry out internal fertilization. Usually they produce fewer zygotes and provide more parental care.

Although there are exceptions, Table 13.1 gives you the general idea about types of fertilization and development.

TABLE 13.1 Types of Fertilization and Development

Animal	Fertilization	Development	Number of Eggs	Parental Care
Fish	External	External	Many	None
Amphibians	External	External	Many	None
Reptiles	Internal	External (inside the egg)	Few	Some
Birds	Internal	External (inside the egg)	Few	Much
Mammals	Internal	Internal	Few	Much

Asexual Reproduction

Asexual reproduction produces offspring genetically identical to the parent. It has several advantages over sexual reproduction.

1. It enables animals living in isolation to reproduce without a mate.

2. It creates numerous offspring quickly.

3. There is no expenditure of energy maintaining elaborate reproductive systems or hormonal cycles.

4. Because offspring are clones of the parent, asexual reproduction is advantageous when the environment is stable and favorable.

Types of Asexual Reproduction in Sample Organisms

- **Fission** is the separation of an organism into two new cells. (Amoebas, bacteria)
- **Budding** involves the splitting off of new individuals from existing ones. (Hydra)
- **Fragmentation** and regeneration occur when a single parent breaks into parts that regenerate into new individuals. (Sponges, planaria, sea stars)
- **Parthenogenesis** involves the development of an egg without fertilization. The resulting adult is haploid. (Honeybees, some lizards)

Sexual Reproduction

Sexual reproduction has one major advantage over asexual reproduction: variation. Each offspring is the product of both parents and may be better able to survive than either parent, especially in an environment that is rapidly changing.

The Human Male Reproductive System

Figure 13.1 illustrates the human male reproductive system.

- **Testes** (testis, singular)—male gonads; the site of sperm formation
- **Vas deferens**—the duct that carries sperm during ejaculation from the epididymis to the penis

- **Prostate gland**—the large gland that secretes semen directly into the urethra
- **Scrotum**—the sac outside the abdominal cavity that holds the testes; the cooler temperature there enables sperm to survive
- **Urethra**—the tube that carries semen (the nutritive fluid that carries sperm) and urine

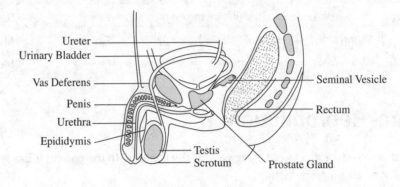

FIGURE 13.1

The Human Female Reproductive System

Figure 13.2 illustrates the human female reproductive system.

- **Ovary**—where meiosis occurs and where the secondary oocyte forms prior to birth
- **Oviduct or fallopian tube**—where fertilization occurs; after ovulation, the egg moves through the oviduct to the uterus
- **Uterus**—where the blastula stage of the embryo implants and develops during the nine-month gestation, should fertilization occur
- **Vagina**—the birth canal; during labor and delivery, the baby passes through the cervix and into the vagina
- **Cervix**—the mouth of the uterus
- **Endometrium**—lining of the uterus

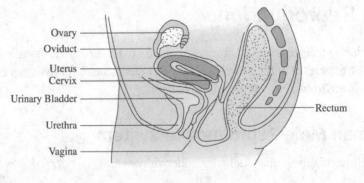

FIGURE 13.2

The Menstrual Cycle

The menstrual cycle consists of a series of changes in the ovary and uterus that is controlled by the interaction of hormones. Human females release a gamete at intervals that average about every 28 days from puberty until menopause. The release of an egg (really a secondary oocyte) is one of four stages of the cycle. See Figure 13.3.

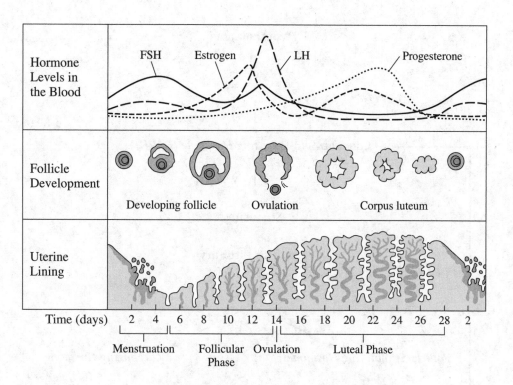

FIGURE 13.3 The menstrual cycle

Follicular Phase

Several tiny cavities called follicles in the ovaries grow and secrete increasing amounts of estrogen in response to follicle-stimulating hormone (FSH) from the anterior pituitary.

Ovulation

The secondary oocyte ruptures out of the ovaries in response to a rapid increase in luteinizing hormone (LH) from the anterior pituitary. Ovulation occurs on or about the 14th day after menstruation.

Luteal Phase

After ovulation, the corpus luteum (the cavity of the follicle left behind) forms and secretes estrogen and progesterone that thicken the endometrium (lining) of the uterus.

Menstruation

If implantation of an embryo does not occur, the buildup of the lining of the uterus breaks down and is shed. Tissue and some blood are discharged from the vagina. This bleeding is commonly called the period.

Hormonal Control of the Menstrual Cycle

The **hypothalamus** in the brain releases GnRH, which stimulates the anterior pituitary to release FSH and LH, which, in turn, stimulate the ovary to release estrogen and progesterone. These two hormones prepare the uterus for implantation of an embryo by causing the uterine lining to thicken in preparation for implantation of an embryo.

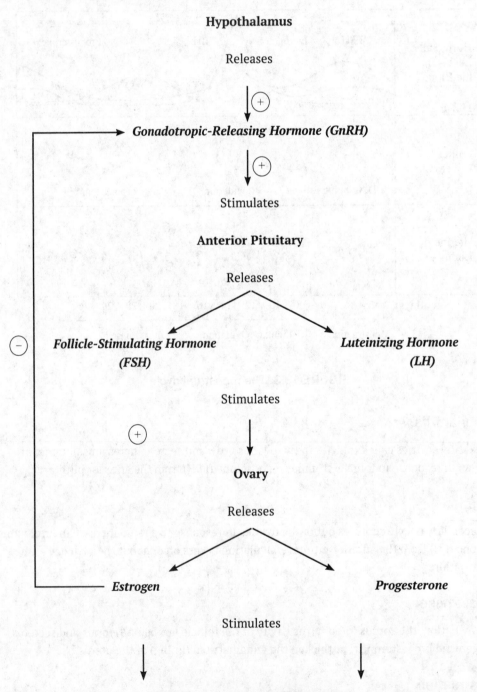

Hypothalamus

Releases

(+)

Gonadotropic-Releasing Hormone (GnRH)

(+)

Stimulates

Anterior Pituitary

Releases

Follicle-Stimulating Hormone (FSH)   *Luteinizing Hormone (LH)*

(−)   (+)

Stimulates

Ovary

Releases

Estrogen   *Progesterone*

Stimulates

Thickens the lining of the uterus

Spermatogenesis

Spermatogenesis, the process of sperm production, is a continuous process that begins at puberty and can continue into old age. It begins as luteinizing hormone (LH) induces the testes to produce testosterone. Together FSH and testosterone stimulate sperm production in the testes.

Each spermatogonium cell (2*n*) divides by *mitosis* to produce two primary spermatocytes (2*n*) that can each undergo *meiosis I* to produce two secondary spermatocytes (*n*). Each secondary spermatocyte then undergoes *meiosis II*, which yields four spermatids (*n*). These spermatids differentiate and move to the epididymis where they become motile. Note that each spermatogonium cell undergoes meiosis to produce four active, equal-sized sperm. Figure 13.4 shows the development of sperm by meiosis.

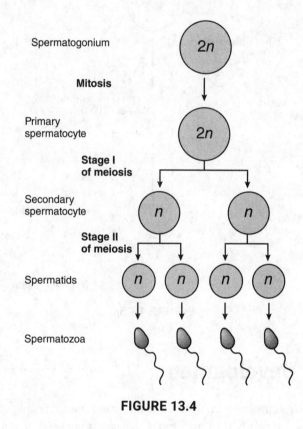

FIGURE 13.4

Oogenesis

Oogenesis, the production of ova, begins prior to birth. A female baby is born with all the primary oocytes she will ever have. Within the embryo, an oogonium cell (2*n*) undergoes *mitosis* to produce two primary oocytes (2*n*). These remain inactive within follicles in the ovaries until puberty, when they become reactivated by hormones. At that time, *meiosis I* occurs, producing **secondary oocytes** (*n*) that are released monthly at ovulation. *Meiosis II* does not occur until a sperm penetrates the secondary oocyte during fertilization. This could be 40 years after meiosis I.

During meiosis I and meiosis II, the cytoplasm divides unequally. This unequal distribution of cellular contents forms one large daughter cell containing most of the organelles and cytoplasm. This cell will develop into the mature ova that can be fertilized. The remaining three smaller cells, called polar bodies, may disintegrate by a process called apoptosis, or remain as support structures while the ova matures. Note that one primary oogonium cell produces only one active egg cell. Figure 13.5 shows the development of an ovum by meiosis.

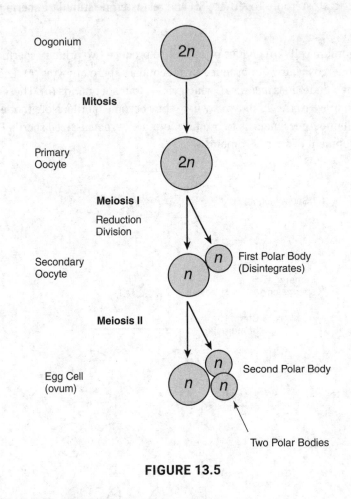

FIGURE 13.5

Embryonic Development

A small, flagellated, haploid sperm (*n*) fertilizes a larger, nonmotile, haploid egg (*n*) to form a diploid (2*n*) zygote. The zygote then undergoes cleavage, a succession of mitotic divisions that results in the formation of a hollow ball called a blastula.

Embryonic development consists of three stages: cleavage, gastrulation, and organogenesis.

Cleavage is the rapid mitotic cell division of the zygote that begins immediately after fertilization. The cells are dividing so quickly that individual cells have no time to grow in size. Embryologists arbitrarily consider the *end of cleavage to be characterized by the production of a fluid-filled ball of cells* called a **blastula**. The individual cells of the blastula are called *blastomeres*, and the fluid-filled center is a *blastocoel*.

Gastrulation is the continuation of the process that began during cleavage. It involves differentiation: the rearrangement of the blastula to produce a *three-layered embryo* called a **gastrula**. The gastrula consists of three differentiated layers called embryonic germ layers. These three germ layers are the **ectoderm**, **endoderm**, and **mesoderm**. They will develop into all the parts of the adult animal. See Figure 13.6.

- The **ectoderm** will become the skin and the nervous system.

- The **endoderm** will form the viscera, including the lungs, liver, and digestive organs.

- The **mesoderm** will give rise to the muscle, blood, and bones. Some primitive animals (sponges and cnidarians) develop a noncellular layer, the **mesoglea**, instead of the mesoderm.

Organogenesis is the process by which cells continue to differentiate, producing organs from the three embryonic germ layers. Once all the organ systems have been developed, the embryo simply increases in size and becomes a fetus.

Overall the pattern of embryo development is:

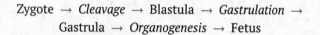

Zygote → *Cleavage* → Blastula → *Gastrulation* →
Gastrula → *Organogenesis* → Fetus

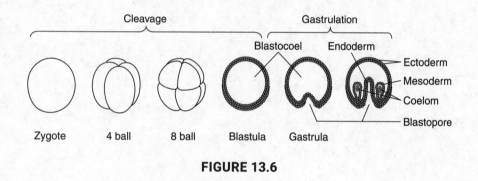

FIGURE 13.6

Extraembryonic Membranes of the Bird Embryo

The four membranes that arise outside the bird embryo are known as the extraembryonic membranes. They consist of the chorion, yolk sac, amnion, and allantois.

1. **Chorion.** It lies under the shell and allows for diffusion of respiratory gases between the outside environment and the inside of the shell.

2. **Yolk sac.** It encloses the yolk, the food for the growing embryo.

3. **Amnion.** It encloses the embryo in protective amniotic fluid.

4. **Allantois.** It is analogous to the placenta in mammals. It is the conduit for respiratory gases to and from the embryo. It is also the place where the nitrogenous waste uric acid accumulates until the chick hatches.

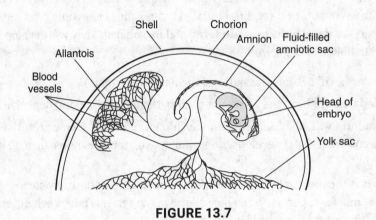

Figure 13.7 illustrates the bird embryo and extraembryonic membranes within the egg.

FIGURE 13.7

The Human Immune System

Learning Objectives

In this chapter, you will learn:

- Nonspecific defense
- Specific defense—third line of defense
- Types of immunity
- Allergies, antibiotics, vaccines, and autoimmune disease

We live in a sea of germs. They are in the air we breathe, the food we eat, and the water we drink and swim in. Diseases are spread in a number of different ways. Some **pathogens** (organisms that cause disease) can be transmitted between people through contact with infected body fluids or respiratory droplets, such as COVID-19 or syphilis. Some, like amoebic dysentery or *E. coli*, occur from the ingestion of contaminated food or water. Still others, like bubonic plague and Zika virus, occur when a vector, such as a flea or mosquito, injects the pathogen into a host.

The human body has evolved a complex system of defenses to fight pathogens and keep us healthy. Some of these defenses are like walls of a fortress. Others are like soldiers that engage in hand-to-hand combat once the invaders have crossed the moat and entered the fortress. There are also cells that act like warriors, shooting projectiles at invaders that have breached the walls of the fortress.

One way in which the immune system defends the body is in a nonspecific way, by attacking anything foreign. The immune system can also identify certain invading cells and attack them specifically. Here are the basics of the immune system.

Nonspecific Defense

The nonspecific immune system consists of two lines of defense.

First Line of Defense

The first line of nonspecific defense is a barrier that helps prevent pathogens from entering the body. The body has several different types of barriers:

- Skin that blocks pathogens
- Mucous membranes that release mucus to trap microbes
- Cilia in the respiratory system that sweep out mucus with its trapped microbes
- Stomach acid that kills germs that enter through the nose and mouth

Second Line of Defense

Microbes that get into the body encounter the second line of nonspecific defense. It is meant to limit the spread of invaders in advance of specific immune responses. There are three types.

1. **Inflammatory response**. It is characterized by swelling, redness, soreness, and increased warmth in the area. The purpose of this process is to increase the blood supply to the area, thus increasing nutrients—including oxygen and white blood cells to fight disease. The inflammatory response works in several ways.

    - **Histamine** triggers vasodilation (enlargement of blood vessels), which increases blood supply to the area, bringing more phagocytes to gobble up germs. Histamine is also responsible for the symptoms of the common cold: sneezing, coughing, redness, and itching and runny nose and eyes—all an attempt to rid the body of invaders.

    - Increased body temperature speeds up the immune system and makes it more difficult for microbes to function.

2. **Phagocytes**. These gobble up invading microbes. Macrophages ("giant eaters") are a type of white blood cell that extend **pseudopods** and engulf huge numbers of microbes over a long period of time.

3. **Interferons**. These chemicals are released by the immune system to block against viral infections.

Specific Defense—Third Line of Defense

The third line of defense is specific and consists of lymphocytes. There are two types of lymphocytes, B lymphocytes (B cells) and T lymphocytes (T cells). Both originate in the bone marrow. Once mature, both cell types circulate in the blood, lymph, and lymphatic tissue: spleen, lymph nodes, tonsils, and adenoids. Both recognize different specific **antigens** (germs). Strictly speaking, an antigen is anything that triggers an immune response.

1. **B lymphocytes**. These produce **antibodies** against a specific antigen in what is called a *humoral response*.

2. **T lymphocytes**. These cells actively find and destroy infected cells. They also signal other immune cells and immune responses in what is called a *cell-mediated response*.

Antibodies

Antibodies are part of the third line of defense—the specific immune response. Each antibody has the ability to bind to only one particular antigen. For example, antibodies against influenza bind to and neutralize only the influenza virus; they have no effect on the polio virus. Antibodies neutralize antigens by binding to them and by forming an antigen-antibody complex that can then be gobbled up by a phagocyte.

Clonal Selection

Clonal selection is a *fundamental mechanism* in the development of immunity. Antigens that have entered the body bind to and activate specific B or T lymphocytes. Once a lymphocyte has been selected, it becomes very metabolically active, proliferates (clones thousands of copies of itself), and differentiates into **plasma cells** and **memory cells**.

Plasma Cells

Plasma cells neutralize antigens immediately in what is called the **primary immune response**. They do not live long.

Memory Cells

Memory cells neutralize the same antigens that plasma cells do, but they remain circulating in the blood in small numbers for a lifetime. You have memory cells circulating in your blood that are specific for every viral infection you have ever been ill with and every disease against which you have been vaccinated. You have memory cells specific for mumps, measles, rubella, polio, and so on. The capacity of the immune system to generate a secondary immune response is called **immunological memory**. The immunological memory is the mechanism that prevents you from getting any specific viral infection, such as chicken pox, more than once.

Types of Immunity

Passive immunity is temporary.

- Antibodies are borrowed and do not survive for long.
- Examples are maternal antibodies that pass through the placenta to the developing fetus or that pass through breast milk to the baby. The first milk that a newborn receives from its mother is called colostrum and is 100 percent antibodies.

Active immunity is permanent.

- You make the antibodies yourself.
- An individual makes his or her own antibodies after being ill and recovering or after being given an immunization or vaccine. A vaccine contains either dead or live viruses or enough of the outer coat of a virus to stimulate a full immune response and to impart lifelong immunity.

ABO Blood Types

ABO antibodies circulate in the plasma of the blood and bind with ABO antigens in the event of an improper transfusion. Certain danger from a transfusion comes when *the recipient has antibodies to the donor's antigens*. However, before someone receives a transfusion of blood, samples of the recipient's and the donor's blood must be mixed in the lab to determine and ensure compatibility. This is called a **cross-match**.

Blood type O is known as the universal donor because it has no blood cell antigens to be clumped by the recipient's blood. Blood type AB is known as the universal recipient because there are no antibodies to clump the donor's blood. Table 14.1 shows the antigens and antibodies present in each blood type.

There are four different human blood types: A, B, O, and AB. Blood types are inherited, and your blood type does not change during your lifetime.

TABLE 14.1 Blood Types

Blood Type	Antigens Present on the Surface of the RBCs	Antibodies Present Circulating in the Plasma
A	A	B (antibodies against B)
B	B	A (antibodies against A)
O	None	A and B (antibodies against A and B)
AB	A and B	No antibodies against A or B

HIV

AIDS stands for acquired immune deficiency syndrome. People with AIDS are highly susceptible to opportunistic diseases, infections, and cancers that take advantage of a collapsed immune system. The virus that causes AIDS, HIV (human immunodeficiency virus), mainly attacks helper T cells. HIV is a retrovirus. Once inside a cell, it transcribes itself in reverse. That means that the viral RNA uses the enzyme reverse transcriptase to make DNA. This is the opposite of the typical DNA transcribing mRNA. The host cell then integrates this newly formed DNA into its own genome.

Allergies, Antibiotics, Vaccines, and Autoimmune Diseases

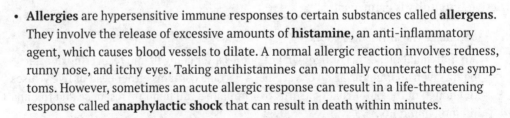

REMEMBER
Antibiotics cure.

- **Allergies** are hypersensitive immune responses to certain substances called **allergens**. They involve the release of excessive amounts of **histamine**, an anti-inflammatory agent, which causes blood vessels to dilate. A normal allergic reaction involves redness, runny nose, and itchy eyes. Taking antihistamines can normally counteract these symptoms. However, sometimes an acute allergic response can result in a life-threatening response called **anaphylactic shock** that can result in death within minutes.

- **Antibiotics** are medicines that kill bacteria or fungi. While vaccines are given to prevent illness caused by viruses, antibiotics are administered *after* a person is sick. They cure the disease.

- **Vaccines prevent** viral infections. There is no cure for viral infections as there is for bacterial infections. New vaccines and treatments are always being developed. In fact, a new vaccine against the human papilloma virus (HPV) that causes cervical cancer is now available. It is the first vaccine designed to prevent cancer. It is recommended for girls between the ages of 11 and 12.

- **Autoimmune diseases** such as multiple sclerosis, lupus, rheumatoid arthritis, and Type 1 diabetes are caused by a terrible mistake of the immune system. The system cannot properly distinguish between self and nonself. Instead, it perceives certain structures in the body as foreign and produces antibodies to attack them. In the case of multiple sclerosis, the immune system attacks the myelin sheath surrounding certain neurons. Treatment for these varies with the disease. In the case of Type 1 diabetes, the treatment is shots of insulin. In other diseases, the goal is to reduce symptoms by suppressing the immune system while maintaining the body's ability to fight disease.

REMEMBER
Vaccines prevent.

Animal Behavior

Behavior is defined as the way an organism responds to changes in its internal or external environment. A **behavior** can be **innate** (inborn), such as running for shelter upon hearing a loud noise. It can also be **learned**, such as one child sharing her toys with another child. Usually behaviors are carried out in response to a **stimulus**, a change in the environment. A monkey eats a banana (behavior) in **response** to hunger (stimulus).

An organism's behavior is important for its survival and for the successful production of offspring. The study of behavior and its relationship to its evolutionary origins is called ethology. Foremost in the field of ethology are three scientists who shared the Nobel Prize in 1973: **Karl von Frisch**, **Konrad Lorenz**, and **Niko Tinbergen**. Von Frisch is known for his extensive studies of honeybee communication and his famous description of the bee waggle dance. Niko Tinbergen is known for his elucidation of the **fixed action pattern**, and Konrad Lorenz is famous for his work with **imprinting**. Here are some basic concepts in the field of animal behavior.

Fixed Action Pattern

A fixed action pattern (FAP) is an innate, *highly stereotypical behavior* that, once begun, is continued to completion no matter how useless or silly looking. FAPs are initiated by external stimuli called **sign stimuli**. When these stimuli are exchanged between members of the same species, they are known as **releasers**. An example of a FAP studied by Tinbergen involves the stickleback fish, which attacks other males that invade its territory. The releaser for the attack is the red belly of the intruder. The stickleback will not attack an invading male stickleback lacking a red underbelly, but it will readily attack a nonfishlike wooden model as long as a splash of red is visible.

Learning

Learning is a sophisticated process in which the responses of the organism are modified as a result of experience. The capacity to learn can be tied to length of life and complexity of the brain. If the animal has a very short life span, like a fruit fly, it has no time to learn, even if it has the ability. It must therefore rely on fixed action patterns. In contrast, if the animal lives a long time and has a complex brain, then a large part of its behavior is dependent on prior experience and learning.

Habituation

Habituation is one of the simplest forms of learning in which an animal comes to ignore a persistent stimulus so it can go about its business. If you tap the dish containing a hydra, it will quickly shrink and become immobile. If you keep tapping, after a while the hydra will begin to ignore the tapping, elongate, and continue moving about. It has become habituated or used to the stimulus.

Associative Learning

Associative learning is one type of learning in which one stimulus becomes linked to another through experience. Examples of associative learning are **classical conditioning** and **operant conditioning**.

Classical Conditioning

Classical conditioning, a type of associative learning, is widely accepted because of the ingenious work of Ivan Pavlov in the 1920s. Normally, dogs salivate when exposed to food. Pavlov trained dogs to associate the sound of a bell with food. The result of this conditioning was that dogs would salivate, an automatic response, upon merely hearing the sound of the bell even though no food was present.

Operant Conditioning

Operant conditioning, also called **trial and error learning**, is another type of associative learning. An animal learns to associate one of its own behaviors with a reward or punishment and then repeats or avoids that behavior. The best-known studies involving operant conditioning were done by B. F. Skinner in the 1930s. In one study, a rat was placed into a cage containing a lever that released a pellet of food. At first, the rat would depress the lever only by accident and would receive food as a reward. The rat soon learned to associate the lever with the food and would depress the lever at will. Similarly, an animal can learn to carry out a behavior to avoid punishment. Such systems of rewards and punishment are the basis of most animal training.

Imprinting

Imprinting is learning that occurs during a sensitive or critical period in the early life of an individual and is irreversible for the length of that period. When you see ducklings following closely behind their mother, you are seeing the result of successful imprinting. Mother-offspring bonding in animals that depend on parental care is critical to the safety and development of the offspring. If the pair does not bond, the parent will not care for the offspring and the offspring will die. At the end of the juvenile period, when the offspring can survive without the parent, the response disappears.

Classic imprinting experiments were carried out by **Konrad Lorenz** with geese. Geese hatchlings will follow the first thing they see that moves. Although the object is usually the mother goose, it can be a box tied to a string or in the case of the classic experiment, it was Konrad Lorenz himself. Lorenz was the first thing the hatchlings saw, and they became imprinted on the scientist. Wherever he went, they followed.

Social Behavior

Social behavior is any kind of interaction between two or more animals, usually of the same species. It is a relatively new field of study, only developed in the 1960s. Types of social behaviors are cooperation, agonistic behavior, dominance hierarchies, territoriality, and altruism.

Cooperation

Cooperation enables the individuals to carry out a behavior, such as hunting, which they can do as a group more successfully than they can do separately. Lions or wild dogs hunt in a pack, enabling them to bring down an animal larger than an individual could ever bring down alone.

Agonistic Behavior

Agonistic behavior is aggressive behavior. It involves a variety of threats or actual combat to settle disputes between individuals. These disputes are commonly over access to food, mates, or shelter. It involves both real aggressive behavior as well as ritualistic or symbolic behavior. One combatant does not have to kill the other. The use of symbolic behavior often prevents serious harm. A dog shows aggression by baring its teeth and erecting its ears and hair. It stands upright to appear taller and looks directly at its opponent. If the aggressor succeeds in scaring the opponent, the loser engages in submissive behavior that says, "You win, I give up." Examples of submissive behaviors are looking down or away from the winner. Submissive dogs or wolves put their tail between their legs and run off. Once two individuals have settled a dispute and established their relationship by agonistic behavior, future encounters between them usually do not involve combat or posturing.

Dominance Hierarchies

Dominance hierarchies are pecking order behaviors that dictate the social position of an animal in a culture. This is commonly seen in hens where the alpha animal (top ranked) controls the behaviors of all the others. The next in line, the beta animal, controls all others except the alpha animal. Each animal threatens all animals beneath it in the **pecking order**. The top-ranked animal is assured of first choice of any resource, including food after a kill, the best territory, or the most-fit mate.

Territoriality

A **territory** is an area an organism defends and from which other members of the community are excluded. Territories are established and defended by agonistic behaviors. They are used for capturing food, mating, and rearing young. The size of the territory varies with its function and the amount of resources available.

Altruism

Altruism is a behavior that reduces an individual's reproductive fitness (the animal may die) while increasing the fitness of the group or family. When a worker honeybee stings an intruder in defense of the hive, the worker bee usually dies. However, the worker bee's action increases the fitness of the queen bee that lays all the eggs. How can altruism evolve if the altruistic individual dies? The answer is called **kin selection**. When an individual sacrifices itself for the family, it is sacrificing itself for relatives (the kin), which share similar genes. The kin are selected as the recipients of the altruistic behavior. They are saved and can pass on their genes. Altruism evolved because it increases the number of copies of a gene common to a related group.

Ecology

> ## Learning Objectives
>
> In this chapter, you will learn:
> - Properties of populations
> - Population growth
> - Community structure and population interactions
> - The food chain
> - Ecological succession
> - Biomes
> - Chemical cycles
> - Humans and the biosphere

Ecology is the study of the interactions of organisms with their physical environment and with each other. Here is some basic vocabulary for the topic.

1. A **population** is a group of individuals of one species living in one area who can interbreed and interact with each other.
2. A **community** consists of all the organisms living in one area.
3. An **ecosystem** includes all the organisms in a given area as well as the abiotic (nonliving) factors with which they interact.
4. **Abiotic factors** are nonliving and include temperature, water, sunlight, wind, rocks, and soil.
5. **Biotic factors** include all the organisms with which an organism might react, such as birds, insects, plants, predators, prey, and parasites.
6. The **biosphere** is the global ecosystem.
7. A **niche** includes what an organism eats and what it needs to survive.

Properties of Populations

Populations are defined by their size, density, and dispersion.

Size

Size is the total number of individuals in a population. Four variables limit the size of a population: the number of births, the number of deaths, immigration, and emigration.

Density

Density is the number of individuals per unit area or volume. It is often very difficult, if not impossible, to count the number of organisms inhabiting a certain area. For example, imagine trying to count the number of ants in 1 acre of land. Instead, scientists use sampling techniques to estimate the number of organisms living in one area. One sampling technique commonly used to estimate the size of a population is called *mark and recapture*. In this technique, organisms are captured, tagged, and then released. Some time later, the same process is repeated and a special mathematical formula is used to determine the density of the population.

Dispersion

Dispersion is the pattern of spacing of individuals within the area the population inhabits. The most common pattern of dispersion is **clumped**. Fish travel this way in schools because there is safety in numbers. Some populations are spread in a **uniform pattern**. For example, certain plants may secrete toxins that keep away other plants that would compete for limited resources. **Random spacing** occurs in the absence of any special attractions or repulsions. Trees can be spaced randomly in a forest. Figure 16.1 shows different dispersion patterns.

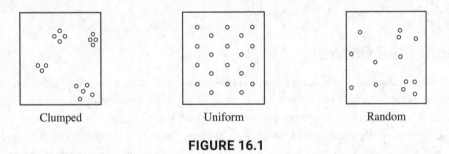

| Clumped | Uniform | Random |

FIGURE 16.1

Population Growth

Every population has a characteristic **biotic potential**, *the maximum rate at which a population could increase under ideal conditions*. Different populations have different biotic potentials that are influenced by several factors. These factors include:

- Age at which reproduction begins
- Life span during which the organisms are capable of reproducing
- Number of reproductive periods in the lifetime
- Number of offspring the organism is capable of having at one time

Regardless of whether a population has a large or small biotic potential, certain characteristics about growth are common to all organisms.

Figure 16.2 shows a graph of classic population growth. After an initial period of slow growth when the organism becomes accustomed to the new environment, the population explodes and grows exponentially. The population grows until it reaches the maximum that the environment can support (**carrying capacity**). After some undetermined amount of time, the population may crash. Many factors can cause a population to crash: predation, parasitism, severe competition, an end to resources, and/or too much waste that poisons the environment.

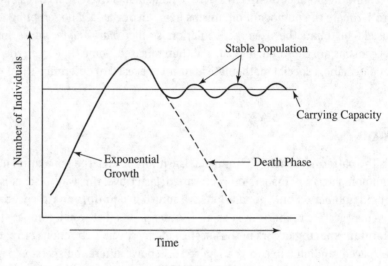

FIGURE 16.2

Exponential Growth

The simplest model for population growth is one with unrestrained or exponential growth—a population with no predation, parasitism, or competition; with no immigration or emigration; and in an environment with unlimited resources. This is characteristic of a population that has been recently introduced into an area, such as a sample of bacteria newly inoculated onto a petri dish. Although exponential growth is usually short-lived in nature, the human population has been in the exponential growth phase for over 300 years.

Carrying Capacity

Ultimately, there is a limit to the number of individuals that can occupy one area at a particular time. That limit is called the carrying capacity (K). Each particular environment has its own carrying capacity around which the population size oscillates.

In addition, the carrying capacity can change as the environmental conditions change. This is illustrated in Figure 16.3.

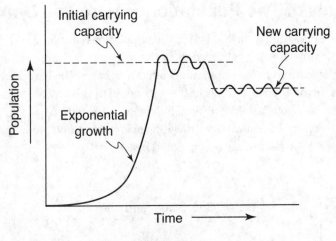

FIGURE 16.3

Reproductive Strategies

Some organisms are opportunistic. They reproduce rapidly when the environment is uncrowded and resources are vast. They are referred to as r-strategists. Insects are good examples.

Other organisms, the K-strategists (K for carrying capacity), tend to maximize population size near the carrying capacity for an environment. Mammals are good examples of K-strategists. Table 16.1 compares r-strategists and K-strategists.

TABLE 16.1 Comparison of Reproductive Strategies

r-strategists	K-strategists
Many, small young	Few, large young
Little or no parenting	Intensive parenting
Rapid maturation	Slow maturation
Reproduce once	Reproduce many times
Example: insects	Example: mammals

Limiting Factors

Limiting factors are those factors that limit population growth. They are divided into two categories: density-dependent and density-independent factors.

- **Density-dependent factors** are those factors that increase directly as the population density increases. They include competition for food, buildup of wastes, predation, and disease.
- **Density-independent factors** are those factors whose occurrence is unrelated to the population density. They include earthquakes, storms, and naturally occurring fires and floods.

A Case Study of Two Populations—Hare and Lynx

A perfect study in population growth involves the populations of snowshoe hare and lynx at the Hudson Bay Company, which kept records of the pelts sold by trappers from 1850 to 1930. The data reveal fluctuations in the populations of both animals. The hare feeds on grass, and the lynx feeds on the hare. So, the cycles in the lynx population are probably caused by cyclic fluctuations in the hare population. Additionally, cycles in the hare population are probably due to a limited food supply, cyclical overcrowding, overgrazing, and predation by the lynx. Figure 16.4 compares the sizes of the two populations.

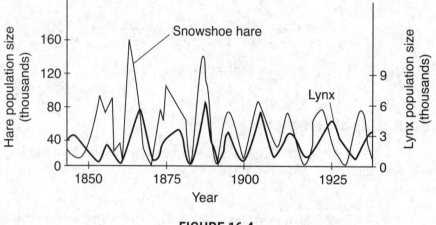

FIGURE 16.4

Community Structure and Population Interactions

Communities are made of populations that interact with the environment and with each other. These interactions are very complex but can be divided into five categories: **competition**, **predation**, **parasitism**, **mutualism**, and **commensalism**.

Competition

The Russian scientist G. F. Gause developed the **competitive exclusion principle** after studying the effects of competition between species in a laboratory setting. He worked with two very similar species, *Paramecium caudatum and Paramecium aurelia*. When he cultured them separately, each population grew rapidly and then leveled off at the carrying capacity. When he put the two cultures together, *P. aurelia* had the advantage and drove the other species to extinction. His principle states that *two species cannot coexist in a community if they share a niche—that is, if they compete for the same resources.*

In nature, there are two possible outcomes besides extinction if two species inhabit the same area and occupy the same niche. One of the species will become extinct as happened in Gause's experiment, or one will evolve through natural selection to exploit different resources. This second process is called **resource partitioning**. Another possibility is what occurred in the Galapagos Islands, where finches evolved different beak sizes and were able to eat different kinds of seeds and avoid competition. This divergence in adaptation is called **character displacement**.

Predation

Predation can refer to one animal eating another animal, or it can also refer to animals eating plants. For their protection, animals and plants have evolved defenses against predation.

Plants have evolved spines, thorns, and chemical poisons such as strychnine, mescaline, morphine, and nicotine to fend off attack by animals.

Animals have evolved active defenses, such as hiding, fleeing, or defending themselves. These, however, can be very costly in terms of energy. Animals have also evolved passive defenses that rely on cryptic coloration or camouflage.

1. **Aposematic coloration**. The very bright, often red or orange coloration of poisonous animals is a warning that possible predators should avoid them.
2. **Batesian mimicry**. This is copycat coloration, where one harmless animal mimics the coloration of another that is poisonous. One example is the viceroy butterfly that is harmless but looks very similar to the monarch butterfly that stores poisons in its body that it absorbs from the milkweed plant.
3. **Müllerian mimicry**. Two or more poisonous species resemble each other and gain an advantage from their combined numbers. Predators learn more quickly to avoid any prey with that appearance.

Feeding

Animals have evolved a variety of ways to feed. **Herbivores** feed on plants. **Carnivores** eat other animals. **Detritivores** are animals that feed on plants and animals that have died and decomposed into organic matter called detritus. Here are some other relationships based on feeding behaviors.

Mutualism

Mutualism is a symbiotic relationship where both organisms benefit (+/+). An example is the bacteria that live in the human intestine and produce vitamins for the host.

Commensalism

Commensalism is a symbiotic relationship in which one organism benefits and one is neither helped nor harmed by the other organism (+/o). Barnacles, which are small, sessile crustaceans that attach themselves to the underside of a whale, benefit by gaining access to a variety of food sources as the whale swims into different areas. The whale is unaware of the barnacles.

Parasitism

Parasitism is a symbiotic relationship (+/−) where one organism, the parasite, benefits while the host is harmed. A tapeworm in the human intestine is an example.

The Food Chain

The **food chain** is the pathway along which energy is transferred from one **trophic** or **feeding level** to another. Energy, in the form of food, moves from the producers to the herbivores to the carnivores. Only about 10 percent of the energy stored in any trophic level is converted to organic matter at the next trophic level. This means that if you begin with 1,000 g of plant matter, the food chain can support 100 g of primary consumers (herbivores), 10 g of secondary consumers (carnivores), and only 1 g of tertiary consumers (carnivores). As a result of the loss of energy from one trophic level to the next, food chains never have more than four or five trophic levels. A good model to demonstrate the interaction of the organisms in the food chain and the loss of energy is the food pyramid shown in Figure 16.5.

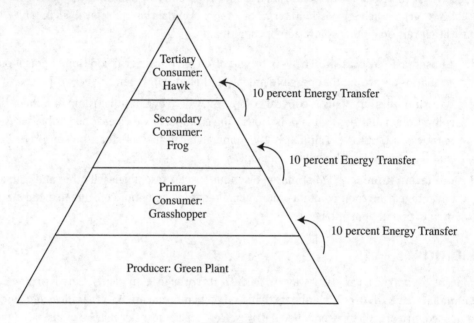

FIGURE 16.5

Food chains are not isolated from their environment. They are interwoven with other food chains into a **food web**. An animal can occupy one trophic level in one food chain and a different trophic level in another food chain. For example, humans can be primary consumers when eating vegetables but are tertiary consumers when eating a steak. Here are two examples of food chains, one terrestrial and one marine, each with four trophic levels.

Producers	→ *Primary Consumers*	→ *Secondary Consumers*	→ *Tertiary Consumers*
		Terrestrial Food Chain	
Green Plant	→ Grasshopper	→ Frog	→ Hawk
		Marine Food Chain	
Phytoplankton	→ Zooplankton	→ Small Fish	→ Shark

Producers

Producers convert light energy to chemical bond energy and have the greatest biomass of any trophic level. They include green plants, **diatoms**, and **phytoplankton**. Diatoms are photosynthetic algae that drift in the oceans. Phytoplankton are algae and photosynthetic bacteria that drift passively in aquatic environments. Diatoms and phytoplankton are the basis for most marine and freshwater aquatic ecosystems.

Primary Consumers

- Eat producers
- Are herbivores
- Examples: grasshoppers and zooplankton (microscopic arthropods)

Secondary Consumers

- Eat primary consumers
- Are carnivores
- Examples: frogs and small fish

Tertiary Consumers

- Eat secondary consumers
- Are carnivores
- Top of the food chain
- Fewer of these (less biomass) than any organism in the food chain
- Least stable trophic level and most sensitive to fluctuations in populations of the other trophic levels
- Examples: hawks and large fish

Energy and Productivity in Food Chains

When ecologists study energy transformations within ecosystems, they discuss **productivity**, which is the rate at which organic matter is created by producers. **Gross primary productivity** is the amount of energy converted to chemical energy by photosynthesis per unit time in an ecosystem. **Net primary productivity** is the gross primary productivity minus the energy used by the primary producers for respiration.

Biological Magnification

Organisms occupying higher trophic levels have greater concentrations of accumulated toxins stored in their bodies than those at lower trophic levels. This is called **biological magnification**. The bald eagle almost became extinct because Americans sprayed crops heavily with the pesticide DDT in the 1950s. DDT entered the bottom of the food chain and accumulated in the bald eagle, at the top of the food chain. Because DDT interferes with the deposition of calcium in eggshells, the thin-shelled eggs were broken easily and few eaglets hatched. DDT is now outlawed, and the bald eagle was saved from extinction.

TIP
Diatoms and phytoplankton are aquatic producers.

Decomposers

Decomposers—bacteria and fungi—are usually not depicted in any diagram of a food chain. However, without decomposers to recycle nutrients back to the soil to nourish plants, there would be no food chain and no life.

Ecological Succession

Most communities are not stable. They are dynamic, always changing. The size of a population increases and decreases around the carrying capacity. Migration of a new species into a habitat can alter the entire food chain. Major disturbances, both natural and human-made, like volcanic eruptions, strip mining, clear-cutting a forest, and forest fires, can suddenly and drastically destroy a community or an entire ecosystem. What follows this destruction is the process of sequential rebuilding of the ecosystem called ecological succession.

If the rebuilding begins in a lifeless area where even soil has been removed, the process is called **primary ecological succession**. *The essential and dominant characteristic of primary ecological succession is soil building.* After an ecosystem is destroyed, the first organisms to inhabit a barren area are **pioneer organisms** like lichens (a symbiont consisting of algae and fungi) and mosses, which are introduced into the area as spores by the wind. Soil develops gradually as rocks weather and organic matter accumulates from the decomposed remains of the pioneer organisms. Once soil is present, pioneer organisms are overrun by other, larger organisms: grasses, bushes, and then trees. The final stable community that remains is called the **climax community**. It remains until the ecosystem is once again destroyed by a **blowout**, a disaster that destroys the ecosystem.

One example of primary succession that was studied in detail is at the southern edge of Lake Michigan. As the glaciers gradually receded northward after the last ice age (10,000 years ago), it left a series of new beaches and sand dunes exposed. Now, someone who begins at the water's edge and walks away from the water for several miles will pass through a series of communities that were formed in the last 10,000 years. These communities represent the various stages, beginning with bare, sandy beach and ending with a climax community of old, well-established forests. In some cases, the climax community is a beech-sugar maple forest. In other areas, the forest is a mix of hickory and oak.

The process known as **secondary succession** occurs when an existing community has been cleared by some disturbance that leaves the soil intact. This can be seen in the 1988 fires in Yellowstone National Park, which destroyed all the old growth that had been dominated by lodge pole pine but left the soil intact. Within one year, the burned areas in Yellowstone were covered with new vegetation.

Biomes

Biomes are very large regions of Earth whose distribution depends on the amount of rainfall and the temperature in an area. Each biome is characterized by different vegetation and animal life. There are many biomes, including freshwater, marine, and terrestrial biomes. In the Northern Hemisphere, from the equator to the most northerly climes, there is a trend in terrestrial biomes: from tropical rain forest, to desert, to grasslands, to temperate deciduous forest, to taiga, and finally to tundra in the north. *Changes in altitude produce effects similar to changes in latitudes.* On the slopes of the Appalachian Mountains in the east and the Rockies and coastal ranges in the west, there is a similar trend in biomes. As elevation increases and temperatures and humidity decrease, one passes through temperate deciduous forest to taiga to tundra. Here is an overview of the major biomes of the world.

Marine

- The largest biome, covering three-fourths of Earth's surface
- The most stable biome, with temperatures that vary little because water has the ability to absorb lots of heat and there is such an enormous volume of water
- Provides most of Earth's food and oxygen
- Subdivided into different regions classified by amount of sunlight they receive, distance from shore and water depth, and whether open water or ocean bottom
- Open oceans are nutrient-poor environments compared with land

Tropical Rain Forest

- This biome is found near the equator with abundant rainfall, stable temperatures, and high humidity.
- Although these forests cover only 4 percent of Earth's land surface, they account for more than 20 percent of Earth's net carbon fixation (food production).
- It has the greatest plant species diversity of any biome on Earth. It may have as many as 50 times the number of species of trees as does a temperate forest.
- Dominant trees are very tall with interlacing tops that form a dense canopy, keeping the floor of the forest dimly lit even at midday. The canopy also prevents rain from falling directly onto the forest floor, but leaves drip rain constantly.
- Many trees are covered with **epiphytes**, photosynthetic plants that grow on other trees rather than supporting themselves. They are not parasites but may kill the trees inadvertently by blocking the light.
- This biome has the most animal species diversity of any biome and includes birds, reptiles, mammals, and amphibians.

Desert

- A desert receives less than 10 inches of rainfall per year; not even grasses can survive.
- A desert experiences the most extreme temperature fluctuations of any biome. Daytime *surface* temperatures can be as high as 70°C. With no moderating influence of vegetation, heat is lost rapidly at night. Shortly after sundown, temperatures drop drastically.
- Characteristic plants are the drought-resistant cacti with shallow roots to capture as much rain as possible during hard and short rains that are characteristic of the desert.
- Other plants include sagebrush, creosote bush, and mesquite.
- There are many small annual plants that are stimulated to grow only after a hard rain. They germinate, send up shoots and flowers, and die all within a few weeks.
- Most animals are active at night or during a brief early-morning period or late afternoon, when the heat is not so intense. During the day, animals remain cool by burrowing underground or hiding in the shade.

TIP
The marine biome is the largest and most stable on Earth.

TIP
The tropical rain forest has the greatest diversity of animal species of any biome.

TIP
Deserts experience the most extreme fluctuations in temperature of any biome.

- Cacti can expand to hold extra water and have modified leaves called spines, which protect against animals attacking a cactus for its water.

- As an example of how severe conditions in a desert can be, in the Sahara desert, there are regions of hundreds of miles across that are completely barren of any vegetation.

- Characteristic animals include rodents, kangaroo rats, snakes, lizards, arachnids, insects, and a few birds.

Figure 16.6 is a sample climate graph to analyze.

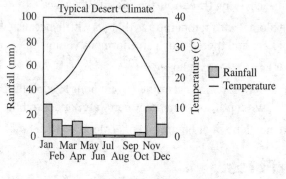

FIGURE 16.6

Temperate Grasslands

- Cover huge areas in both the temperate and tropical regions of the world

- Characterized by low total annual rainfall or uneven seasonal occurrence of rainfall, making conditions inhospitable for forests

- Principal grazing mammals include bison and pronghorn antelope in the United States, and wildebeest and gazelle in Africa

- Burrowing mammals, such as prairie dogs and other rodents, are common

Temperate Deciduous Forest

- Found in the northeast of North America, south of the taiga, and characterized by trees that drop their leaves in winter

- Includes many more plant species than does the taiga

- Shows *vertical stratification* of plants and animals—some species live on the ground, some in the low branches, and some in the treetops

- Rich soil due to decomposition of leaf litter

- Principal mammals include squirrels, deer, foxes, and bears that are dormant or hibernate through the cold winter

Conifer Forest—Taiga or Boreal Forest

- Found in northern Canada and much of the world's northern regions, including Alaska, Russia, and northern Europe

- Dominated by conifer (evergreen) forests, like spruce and fir

- Landscape is dotted with lakes, ponds, and bogs

- Abundance of rainfall allows trees to dominate the landscape

TIP
The taiga—the conifer forest—is the largest terrestrial biome.

- Has very cold winters
- Is the largest terrestrial biome
- Characterized by heavy snowfall; trees are shaped with branches directed downward to prevent heavy accumulations of snow from breaking their branches
- Principal large mammals include moose, black bears, lynx, elk, wolverines, martens, and porcupines
- Flying insects and birds are prevalent in summer
- Has greater variety in animal species than does the tundra
- Seasonal temperature ranges of more than 70°C

Tundra

- Located in the far northern parts of North America, Europe, and Asia
- Called the permafrost, **permanently** frozen subsoil found in the farthest points north, including Alaska
- Commonly referred to as the frozen desert because it gets very little rainfall, which cannot penetrate the frozen ground
- Has the appearance of gently rolling treeless plains with many lakes, ponds, and bogs in depressions
- Insects, particularly flies, are abundant
- Vast numbers of birds nest in the tundra in the summer to eat the insects and migrate south in the winter
- Principal mammals include reindeer, caribou, Arctic wolves, Arctic foxes, Arctic hares, lemmings, and polar bears
- Though the number of individual organisms in the tundra is high, the number of species is small

Chemical Cycles

Although Earth receives a constant supply of energy from the sun, chemicals must be recycled. There are several chemical cycles: the carbon, nitrogen, and water cycles.

The Water Cycle

Water evaporates from Earth, forms clouds, and rains over the oceans and land. Some rain percolates through the soil deeply enough to become groundwater. Some runs along the ground until it enters a river or stream and makes its way back to the seas. Some evaporates directly from the land, but most evaporates from plants by **transpiration**. See Figure 16.7.

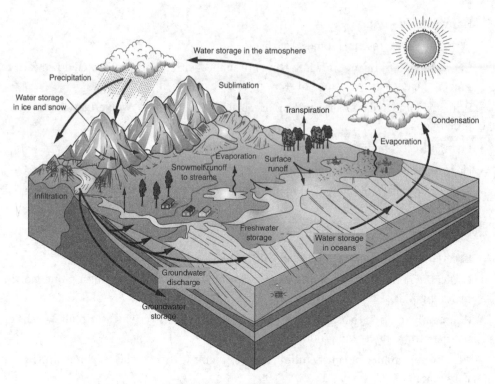

FIGURE 16.7 Water cycle

The Carbon Cycle

The basis of this are the reciprocal processes of photosynthesis and respiration.

- Cell respiration by animals and bacterial decomposers adds carbon dioxide to the air and removes oxygen.
- The burning of fossil fuels adds carbon dioxide to the air.
- Photosynthesis removes carbon dioxide from the air and adds oxygen.

The Nitrogen Cycle

Very little nitrogen enters ecosystems directly from the air. Most of it enters ecosystems by way of bacterial processes. See Figure 16.8.

- **Nitrogen-fixing bacteria** live in the nodules in the roots of legumes and convert free nitrogen (N_2) into the ammonium ion (NH_4^+).
- **Nitrifying bacteria** convert the ammonium ion (NH_4^+) into nitrites (NO_2^-) and then into nitrates (NO_3^-).
- **Denitrifying bacteria** convert nitrates (NO_3^-) into free atmospheric nitrogen (N_2).
- **Decomposers** are bacteria that break down dead organic matter, like dead plants and animals, into the ammonium ion (NH_4^+).

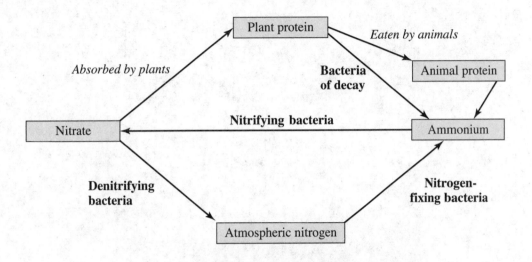

FIGURE 16.8 Nitrogen cycle

Humans and the Biosphere

As the human population has grown in size, we have intruded upon and altered or destroyed many ecosystems. We are responsible for the deforestation of millions of acres of land and the destruction of vast wetlands. We have caused **groundwater contamination** and **depletion**, the **elimination of habitats**, and the **loss of biodiversity**. Humans threaten to make Earth uninhabitable as our population increases exponentially and as we waste natural resources and pollute the air and water. Here are some specific examples of how humans have altered Earth's ecosystem.

Eutrophication of the Lakes

We have disrupted freshwater ecosystems, causing a process called eutrophication. Runoff from sewage, fertilizer, and manure from pastures increases nutrients in lakes and causes excessive growth of algae and other plants. Shallow areas become choked with weeds, and swimming and boating become impossible. As large populations of photosynthetic organisms die, two things happen. First, organic material accumulates on the bottom of the lakes and reduces the depth of the lakes. Second, detritivores use up oxygen as they decompose the dead organic matter. Lower oxygen levels make it impossible for some fish to live. As more fish die, decomposers expand their activity and oxygen levels continue to decrease. The process continues, more organisms die, the oxygen levels decrease, more decomposing matter accumulates on the bottom of the lakes, and, ultimately, the lakes disappear. Figure 16.9 shows the eutrophication of a lake over ten years.

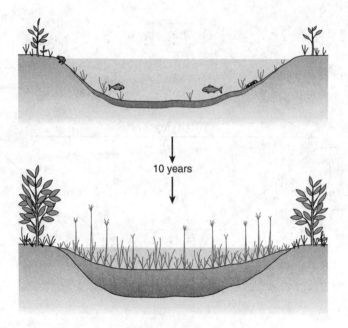

FIGURE 16.9

Acid Rain

Acid rain is caused by pollutants in the air from the combustion of fossil fuels. Nitrogen and sulfur pollutants in the air turn into nitric, nitrous, sulfurous, and sulfuric acids, which cause the pH of the rain to be less than 5.6. This causes the death of organisms in lakes and damages ancient stone architecture.

Toxins

Toxins from industry have gotten into the food chain. Most cattle and chicken feeds contain antibiotics and hormones to accelerate animal growth but may have serious ill effects on humans who eat the chicken and beef. Any carcinogens or teratogens (causing birth defects) that get into the food chain accumulate and remain in our body's fatty tissues.

Global Warming

Excessive burning of fossil fuels has caused the concentrations of carbon dioxide in the air to increase to such high levels that it causes the **greenhouse effect**. This means that carbon dioxide and water vapor in the air absorb much of the infrared radiation reflecting off of Earth, causing the average temperature on Earth to rise. This increase in temperature is called **global warming** and could have disastrous effects for Earth. An increase of 1.0°C on average temperature worldwide would cause the *polar ice caps to melt*, raising the level of the seas and causing tremendous changes in weather patterns, including an increase in occurrences of violent storms. Eventually, major U.S. coastal cities, including New York and Miami, could be under water.

Depleting the Ozone Layer

The accumulations in the air of **chlorofluorocarbons**, chemicals used for refrigerants and aerosol cans, have caused the formation of a hole in the protective **ozone layer**. This allows more ultraviolet (UV) light to reach Earth, which in turn is responsible for an increase in the incidence of **skin cancer** worldwide.

Introduction of New Species

Humans have moved species from one area to another with serious consequences. Two examples are the "killer" honeybees and the zebra mussel.

The "Killer" Honeybees

The African honeybee is a very aggressive subspecies of honeybee that was brought to Brazil in 1956 to breed a variety of bee that would produce more honey in the tropics than the Italian honeybee. The African honeybees escaped by accident and have been spreading throughout the Americas. By the year 2000, these bees killed ten people in the United States.

The Zebra Mussel

In 1988, the zebra mussel, a fingernail-sized mollusk native to Asia, was discovered in a lake near Detroit. No one knows how the mussel got transplanted there. However, scientists assume that it was accidentally carried by a ship from a freshwater port in Europe to the Great Lakes. Without any local natural predator to limit its poplation, the mussel population exploded. They were first discovered when they were found to have clogged the water intake pipes of those cities whose water is supplied by Lake Erie. To date, the zebra mussel has caused millions of dollars of damage. In addition, the influx of the zebra mussel threatens several indigenous species with extinction by outcompeting them.

Pesticides vs. Biological Control

Scientists have developed a variety of pesticides, which are chemicals that kill organisms that we consider to be undesirable. These include insecticides, herbicides, fungicides, and mice and rat killers. On the one hand, these pesticides save lives by increasing food production and by killing animals that carry and cause diseases like bubonic plague (diseased rats) and malaria (anopheles mosquitoes). On the other hand, exposure to pesticides can cause cancer in humans. Moreover, spraying with pesticides ensures the development of resistant strains of pests through natural selection. The pests come back stronger than before. This problem requires that we spray more and more, which means more people will be exposed to these toxic chemicals.

An alternative to widescale spraying with pesticides is called biological control. The following are some biological solutions to get rid of pests without using dangerous chemicals.

1. Use crop rotation—change the crop planted in a field.

2. Introduce natural enemies of the pests—you must be careful, however, that you do not disrupt a delicate ecological balance by introducing an invasive species.

3. Use natural plant toxins instead of synthetic ones.

4. Use insect birth control—male insect pests can be sterilized by exposing them to radiation and then releasing them into the environment to mate unsuccessfully with females.

Index

A

ABO blood types, 164
Abiotic factors, 170
Abscisic acid (ABA), 125
Accessory pigments, 48
Acetylcholine (ACh), 24
Acetyl-CoA, 42
Acid rain, 6, 184
Acidity, 5–6
Acoelomates, 103–104
Actin, 20
Action potential, 145–146
Active immunity, 163
Active transport, 24–27, 27
Adaptive radiation, 92
Adenine (A), 69, 71
Adenosine, 39
Adenosine triphosphate, 39
Adhesion properties, of water, 5
ADP, 40
Adrenaline (epinephrine), 141
Adrenocorticotropic hormone (ACTH), 141
Adventitious root, 116
Aerial roots, 116
Aerobic cell respiration, 41
Aerobic respiration, equation for, 39
Agonistic behavior, 168
AIDS (HIV), 84, 164
Air spaces, 52, 118
Albino fish, 96
Alcohol fermentation, 45–46
Alkalinity, 5–6
Allantois, 150, 159
Alleles, 55
Allergens, 164
Allergies, 164–165
Alternation of generations, 100, 123
Altruism, 169
Alveoli, 22, 136
Alzheimer's disease, 66
Amino acids, 73, 74

Ammonia, 128
Amnion, 159
Amoebas, 99
Amoebic dysentery, 100, 161
Amphibians, 107
Anaerobic cell respiration, 45–46
Anaerobic heterotrophic prokaryotes, 94
Anaerobic phase, 41
Analogous structures, 82
Anaphase, mitosis, 36
Anaphylactic shock, 164
Anatomy, comparative, 81–82
Androgens, 141
Aneuploidy, 64, 77
Angiosperms, 111
Animal, behavior, 166–169
Animal cell, 16-22
Animal starch, 8
Animalia kingdom, 97, 101
Animals
  body temperature regulation, 127
  characteristics to move to land, 94
  Eukarya domain, 99
  evolutionary trends in, 101–104
  excretion, 128
  movement and locomotion, 126–127
  polysaccharides in, 8
Anion, 2
Annelida, 103, 126
Annelids, 101, 106, 126
Anteater, spiny, 107
Antennae, 48
Antheridium, 123
Arthropods, 106
Antibiotics, 84, 164–165
Antibodies, 162–163
Antidiuretic hormone (ADH), 141, 150
Antigens, 162

Apical meristem, 113
Apoplast, 115
Apoptosis, 18
Aposematic coloration, 175
Appalachian Mountains, biome trends, 179
Arachnida, 106
Archaea domain, 97–98, 99
Archaebacteria, 97
Archaeopteryx, 81
Archegonium, 123
Artery, 132, 137, 139
Arthropoda, 101, 103, 126
Arthropods, 101
Asexual reproduction, 121, 153
Associative learning, 167
Athlete's foot, from kingdom Fungi, 100
Atom, 1
Atomic structure, 1–2
ATP (adenosine triphosphate), 39–41, 44
ATP synthase, 21
ATP synthase channels, 43, 44, 49
ATP synthase complexes, 19
Autoimmune diseases, 164–165
Autonomic system, 143
Autosomal dominant, genetic disorders, 65
Autosomal recessive, genetic disorders, 65
Autosomes, 63
Autotrophs, 47, 99–100, 110
Auxins, 124–125
Axons, 144
AZT, 84

B

B lymphocytes, 162
Backcross, 57
Bacteria, as prokaryotic cell, 15
Bacteria domain, 97–99
Bacterial transformation, 67

Bacteriophages, 68
Balanced molecule, 3
Balanced polymorphism, 85
Bald eagle, 177
Baldness, sex influence on pattern, 62
Base-pair substitution, 76
Basic chemistry, atomic structure, 1–2
Batesian mimicry, 175
Bee waggle dance, 166
Behavior, animal, 166–169
Behavioral isolation, 90
Bicarbonate buffering system, 1
Bicarbonate ion (HCO3-), 6
Bilateral symmetry, 102
Bile, 135
Binomial nomenclature, 97
Biochemistry, 1, 82
Biodiversity, loss of, 183
Biogeography, 83
Biological control, vs pesticides, 185–186
Biological magnification, 177
Biology, molecular, 82
Biomes, 178–181
Biosphere, 170, 183–186
Biotechnology, 78
Biotic factors, 170
Biotic potential, 171
Birds, 107
Blastula, 158
Blind albino fish, 96
Blood, 137–139
Blood pressure, 138
Blood types, 164
Body cavity formation, 103–104
Body plan, 129
Body symmetry, in animals, 102
Body temperature regulation, 127
Bonding, 2–3
Boreal forest, 180–181
Bottleneck effect, 86–87
Bowman's capsule, 150